ECOLOGICAL CRISIS

READINGS FOR SURVIVAL

ECOLOGICAL CRISIS

READINGS FOR SURVIVAL

Edited by

Glen A. Love
University of Oregon

and

Rhoda M. Love
Lane Community College

HARCOURT BRACE JOVANOVICH, INC.
NEW YORK • CHICAGO • SAN FRANCISCO • ATLANTA • DALLAS

PHOTOGRAPH ACKNOWLEDGMENTS

Harbrace: 1, 17, 59, 123, 239; © George Knight: 271.

Cover photo: Harbrace.

ISBN: 0-15-518778-3

Library of Congress Catalog Card Number: 77-129475

Printed in the United States of America

To Stan and Jen in the hope of a livable future

PREFACE

"Human history becomes more and more a race between education and catastrophe"—H. G. Wells, *The Outline of History*

The survival of man—and of every living thing on this planet—is threatened. Our air, our water, our land grow more polluted by the day as the technology of production outstrips our ability to preserve an environment hospitable to life. When man participates in the destruction of the world around him, he unwittingly helps to bring about his own destruction.

This book proceeds from the premise that man and the environment he lives in are inextricably linked, and that what affects man's environment is of concern to everyone—scientists, politicians, teachers, and students. No longer is ecology solely the concern of scholars in a special branch of the biological sciences. The study of environmental problems is properly a subject of general education; it is appropriate that people in the humanities and social sciences now find it as important as their colleagues in the natural sciences have in the past.

Because the editors of this volume feel strongly that the ecological crisis involves everyone, we have not confined our selections to the writings of professional scientists. A sound grounding in the basic issues and problems is obviously useful, and such eminent biologists as Sir Julian Huxley, Paul Ehrlich, Marston Bates, Barry Commoner, and Garrett Hardin show that the scientific community has long been actively concerned with what is happening to the environment.

But the crisis in our environment has enormous social consequences, and it is clear that solutions to the crisis will depend largely on the active involvement of all people. Accordingly, essays written by public officials—Senator Gaylord Nelson, Supreme Court Justice William O. Douglas—and scholars and writers from disciplines outside biology—

Aldous Huxley, Kenneth Boulding, Robert Rienow—make up an important part of this book.

Simply to study and reflect upon the problems of our environment is not enough, though; awareness should lead to involvement and action. Concerned citizens—and this most definitely includes students—can voice their concern over the decline in environmental quality by writing letters—to public officials, news media, polluters—on the issues and problems raised here or on problems in their own communities.

As Aldous Huxley points out in the final essay in this volume, the environmental crisis—unlike the problems of nationalism or power politics—*can* respond to rational means of solution. The intent of this book is to provoke discussion—and action—toward an enhanced quality of life for all people on this earth.

GLEN A. LOVE
Associate Professor of English
and Director of Composition
University of Oregon
Eugene, Oregon

RHODA M. LOVE
Instructor in Biology
Lane Community College
Eugene, Oregon

Contents

Prophecy

Here is a nightmarish prediction of what is in store for us within the next ten years if we continue the present destruction of our environment. The scenario seems less like fantasy than recent history, especially as the reader makes his way through this book. Paul Ehrlich (1932-) touches upon a number of issues in this article that receive more exhaustive coverage in later essays by other writers: Ehrlich's fictitious insecticide, Thanodrin, will find a counterpart in the deadly poison, Endrin, described by Frank Graham, Jr., in "Mississippi Fish Kill" (page 153); Ehrlich's prediction of disastrous buildups of air pollution is fulfilled in Berton Roueché's "The Fog" (page 125); his fearful population explosion already has its fuse sizzling as Sir Julian Huxley (page 61) and other population experts warn us. Thus, "Eco-Catastrophe!" is more than harmless fantasy.

Paul Ehrlich, a professor of biology at Stanford University, is very active in the population control movement. He is the author of The Population Bomb *and, with his wife, Anne H. Ehrlich,* Population, Resources, Environment: Issues in Human Ecology.

ECO-CATASTROPHE!

by Paul Ehrlich

The end of the ocean came late in the summer of 1979, and it came even more rapidly than the biologists had expected. There had been signs for more than a decade, commencing with the discovery in 1968 that DDT slows down photosynthesis in marine plant life. It was announced in a short paper in the technical journal, *Science*, but to ecologists it smacked of doomsday. They knew that all life in the sea depends on photosynthesis, the chemical process by which green plants bind the sun's energy and make it available to living things. And they knew that DDT and similar chlorinated hydrocarbons had polluted the entire surface of the earth, including the sea.

But that was only the first of many signs. There had been the final gasp of the whaling industry in 1973, and the end of the Peruvian anchovy fishery in 1975. Indeed, a score of other fisheries had disappeared quietly from over-exploitation and various eco-catastrophes by 1977. The term "eco-catastrophe" was coined by a California ecologist in 1969 to describe the most spectacular of man's attacks on the systems which sustain his life. He drew his inspiration from the Santa Barbara offshore oil disaster of that

---

"Eco-Catastrophe!" From the September 1969 issue of *Ramparts*. Reprinted with the permission of the author.

year, and from the news which spread among naturalists that virtually all of the Golden State's seashore bird life was doomed because of chlorinated hydrocarbon interference with its reproduction. Eco-catastrophes in the sea became increasingly common in the early 1970's. Mysterious "blooms" of previously rare micro-organisms began to appear in offshore waters. Red tides—killer outbreaks of a minute single-celled plant—returned to the Florida Gulf coast and were sometimes accompanied by tides of other exotic hues.

It was clear by 1975 that the entire ecology of the ocean was changing. A few types of phytoplankton were becoming resistant to chlorinated hydrocarbons and were gaining the upper hand. Changes in the phytoplankton community led inevitably to changes in the community of zooplankton, the tiny animals which eat the phytoplankton. These changes were passed on up the chains of life in the ocean to the herring, paice, cod and tuna. As the diversity of life in the ocean diminished, its stability also decreased.

Other changes had taken place by 1975. Most ocean fishes that returned to fresh water to breed, like the salmon, had become extinct, their breeding streams so dammed up and polluted that their powerful homing instinct only resulted in suicide. Many fishes and shellfishes that bred in restricted areas along the coasts followed them as onshore pollution escalated.

By 1977 the annual yield of fish from the sea was down to 30 million metric tons, less than one-half the per capita catch of a decade earlier. This helped malnutrition to escalate sharply in a world where an estimated 50 million people per year were already dying of starvation. The United Nations attempted to get all chlorinated hydrocarbon insecticides banned on a worldwide basis, but the move was defeated by the United States. This opposition was generated primarily by the American petrochemical industry, operating hand in glove with its subsidiary, the United States Department of Agriculture. Together they persuaded the government to oppose the U.N. move—which was not difficult since most Americans believed that Russia and China were more in need of fish products than was the United States. The United Nations also attempted to get fishing nations to adopt strict and enforced catch limits to preserve dwindling stocks. This move was blocked by Russia, who, with the most modern electronic equipment, was in the best position to glean what was left in the sea. It was, curiously, on the very day in 1977 when the Soviet Union announced its refusal that another ominous article appeared in *Science*. It announced that incident solar radiation had been so reduced by worldwide air pollution that serious effects on the world's vegetation could be expected.

Apparently it was a combination of ecosystem destablization, sunlight reduction, and a rapid escalation in chlorinated hydrocarbon pollution from massive Thanodrin applications which triggered the ultimate catastrophe. Seventeen huge Soviet-financed Thanodrin plants were operating in underdeveloped countries by 1978. They had been part of a massive Russian "aid offensive" designed to fill the gap caused by the collapse of America's ballyhooed "Green Revolution."

It became apparent in the early '70s that the "Green Revolution" was more talk than substance. Distribution of high yield "miracle" grain seeds had caused temporary local spurts in agricultural production. Simultaneously, excellent weather had produced record harvests. The combination permitted bureaucrats, especially in the United States Department of Agriculture and the Agency for International Development (AID), to reverse their previous pessimism and indulge in an outburst of optimistic propaganda about staving off famine. They raved about the approaching transformation of agriculture in the underdeveloped countries (UDCs). The reason for the propaganda reversal was never made clear. Most historians agree that a combination of utter ignorance of ecology, a desire to justify past errors, and pressure from agro-industry (which was eager to sell pesticides fertilizers, and farm machinery to the UDCs and agencies helping the UDCs) was behind the campaign. Whatever the motivation, the results were clear. Many concerned people, lacking the expertise to see through the Green Revolution drivel, relaxed. The population-food crisis was "solved."

But reality was not long in showing itself. Local famine persisted in northern India even after good weather brought an end to the ghastly Bihar famine of the mid-'60s. East Pakistan was next, followed by a resurgence of general famine in northern India. Other foci of famine rapidly developed in Indonesia, the Philippines, Malawi, the Congo, Egypt, Colombia, Ecuador, Honduras, the Dominican Republic, and Mexico.

Everywhere hard realities destroyed the illusion of the Green Revolution. Yields dropped as the progressive farmers who had first accepted the new seeds found that their higher yields brought lower prices—effective demand (hunger plus cash) was not sufficient in poor countries to keep prices up. Less progressive farmers, observing this, refused to make the extra effort required to cultivate the "miracle" grains. Transport systems proved inadequate to bring the necessary fertilizer to the fields where the new and extremely fertilizer-sensitive grains were being grown. The same systems were also inadequate to move produce to markets. Fertilizer plants were not built fast enough, and most of the underdeveloped countries could not scrape together funds to purchase supplies, even on concessional terms. Finally, the inevitable happened, and pests began to reduce yields in even the most carefully cultivated fields. Among the first were the famous "miracle rats" which invaded Philippine "miracle rice" fields early in 1969. They were quickly followed by many insects and viruses, thriving on the relatively pest-susceptible new grains, encouraged by the vast and dense plantings, and rapidly acquiring resistance to the chemicals used against them. As chaos spread until even the most obtuse agriculturists and economists realized that the Green Revolution had turned brown, the Russians stepped in.

In retrospect it seems incredible that the Russians, with the American mistakes known to them, could launch an even more incompetent program of aid to the underdeveloped world. Indeed, in the early 1970's there were cynics in the United States who claimed that outdoing the stupidity of American foreign aid would be physically impossible. Those critics were, however, obviously unaware that the Russians had been busily destroying their own environment for many years. The virtual disappearance of sturgeon from Russian rivers caused a great shortage of caviar by 1970. A standard joke among Russian scientists at that time was that they had created an artificial caviar which was indistinguishable from the real thing —except by taste. At any rate the Soviet Union, observing with interest the progressive deterioration of relations between the UDCs and the United States, came up with a solution. It had recently developed what it claimed was the ideal insecticide, a highly lethal chlorinated hydrocarbon complexed with a special agent for penetrating the external skeletal armor of insects. Announcing that the new pesticide, called Thanodrin, would truly produce a Green Revolution, the Soviets entered into negotiations with various UDCs for the construction of massive Thanodrin factories. The USSR would bear all the costs; all it wanted in return were certain trade and military concessions.

It is interesting now, with the perspective of years, to examine in some detail the reasons why the UDCs welcomed the Thanodrin plan with such open arms. Government officials in these countries ignored the protests of their own scientists that Thanodrin would not solve the problems which plagued them. The governments now knew that the basic cause of their problems was overpopulation, and that these problems had been exacerbated by the dullness, daydreaming, and cupidity endemic to all governments. They knew that only population control and limited development aimed primarily at agriculture could have spared them the horrors they now faced. They knew it, but they were not about to admit it. How much easier it was simply to accuse the Americans of failing to give them proper aid; how much simpler to accept the Russian panacea.

And then there was the general worsening of relations between the United States and the UDCs. Many things had contributed to this. The situation in America in the first half of the 1970's deserves our close scrutiny. Being more dependent on imports for raw materials than the Soviet Union, the United States had, in the early 1970's, adopted more and more heavy-handed policies in order to insure continuing supplies. Military adventures in Asia and Latin America had further lessened the international credibility of the United States as a great defender of freedom—an image which had begun to deteriorate rapidly during the pointless and fruitless Viet-Nam conflict. At home, acceptance of the carefully manufactured image lessened dramatically, as even the more romantic and chauvinistic citizens began to understand the role of the military and the industrial system in what John Kenneth Galbraith had aptly named "The New Industrial State."

At home in the USA the early '70s were traumatic times. Racial violence grew and the habitability of the cities diminished, as nothing substantial was done to ameliorate either racial inequities or urban blight. Welfare rolls grew as automation and general technological progress forced more and more people into the category of "unemployable." Simultaneously a taxpayers' revolt occurred. Although there was not enough money to build the schools, roads, water systems, sewage systems, jails, hospitals, urban transit lines, and all the other amenities needed to support a burgeoning population, Americans refused to tax themselves more heavily. Starting in Youngstown, Ohio in 1969 and followed closely by Richmond, California, community after community was forced to close its schools or curtail educational operations for lack of funds. Water supplies, already marginal in quality and quantity in many places by 1970, deteriorated quickly. Water rationing occurred in 1723 municipalities in the summer of 1974, and hepatitis and epidemic dysentery rates climbed about 500 per cent between 1970-1974.

Air pollution continued to be the most obvious manifestation of environmental deterioration. It was, by 1972, quite literally in the eyes of all Americans. The year 1973 saw not only the New York and Los Angeles smog disasters, but also the publication of the Surgeon General's massive report on air pollution and health. The public had been partially prepared for the worst by the publicity given to the U.N. pollution conference held in 1972. Deaths in the late '60s caused by smog were well known to scientists, but the public had ignored them because they mostly involved the early demise of the old and sick rather than people dropping dead on the freeways. But suddenly our citizens were faced with nearly 200,000 corpses and massive documentation that they could be the next to die from respiratory disease. They were not ready for that scale of disaster. After all, the U.N. conference had not predicted that accumulated air pollution would make the planet uninhabitable until almost 1990. The population was terrorized as TV screens became filled with scenes of horror from the disaster areas. Especially vivid was NBC's coverage of hundreds of unattended people choking out their lives outside of New York's hospitals. Terms like nitrogen oxide, acute bronchitis and cardiac arrest began to have real meaning for most Americans.

The ultimate horror was the announcement that chlorinated hydrocarbons were now a major constituent of air pollution in all American cities. Autopsies of smog disaster victims revealed an average chlorinated hydrocarbon load in fatty tissue equivalent to 26 parts per million of DDT. In October, 1973, the Department of Health, Education and Welfare announced studies which showed unequivocally that increasing death rates from hypertension, cirrhosis of the liver, liver cancer and a series of other diseases had resulted from the chlorinated hydrocarbon load. They estimated that Americans born since 1946 (when DDT usage began) now had a life expectancy of only 49 years, and predicted that if current patterns continued, this expectancy would reach 42 years by 1980, when it might level out. Plunging insurance stocks triggered a stock market panic. The president of a major pesticide producer went on television to "publicly eat a teaspoonful of DDT" (it was really powdered milk) and announce that HEW had been infiltrated by Communists. Other giants of the petrochemical industry, attempting to dispute the indisputable evidence, launched a massive pressure campaign on Congress to force HEW to "get out of agriculture's business." They were aided by the agro-chemical journals, which had decades of experience in misleading the public about the benefits and dangers of pesticides. But by now the public realized that it had been duped. The Nobel Prize for medicine and physiology was given to Drs. J. L. Radomski and W. B. Deichmann, who in the late 1960's had pioneered in the documentation of the long-term lethal effects of chlorinated hydrocarbons. A Presidential Commission with unimpeachable credentials directly accused the agro-chemical complex of "condemning many millions of Americans to an early death." The year 1973 was the year in which Americans finally came to understand the direct threat to their existence posed by environmental deterioration.

And 1973 was also the year in which most people finally comprehended the indirect threat. Even the president of Union Oil Company and several other industrialists publicly stated their concern over the reduction of bird populations which had resulted from pollution by DDT and other chlorinated hydrocarbons. Insect populations boomed because they were resistant to most pesticides and had been freed, by the incompetent use of those pesticides, from most of their natural enemies. Rodents swarmed over crops, multiplying rapidly in the absence of predatory birds. The effect of pests on the wheat crop was especially disastrous in the summer of 1973, since that was also the year of the great drought. Most of us can remember the shock which greeted the announcement by atmospheric physicists that the shift of the jet stream which had caused the drought was probably permanent. It signalled the birth of the Midwestern desert. Man's air-polluting activities had by then caused gross changes in climatic patterns. The news, of course, played hell with commodity and stock markets. Food prices skyrocketed, as savings were poured into hoarded canned goods. Official assurances that food supplies would remain ample fell on deaf ears, and even the government showed signs of nervousness when California migrant field workers went out on strike again in protest against the continued use of pesticides by growers. The strike burgeoned into farm burning and riots. The workers, calling themselves "The Walking Dead," demanded immediate compensation for their shortened lives, and crash research programs to attempt to lengthen them.

It was in the same speech in which President Edward Kennedy, after much delay, finally declared a national emergency and called out the National Guard to harvest California's crops, that the first mention of population control was made. Kennedy pointed out that the United States would no longer be able to offer any food aid to other nations and was likely to suffer food shortages herself. He suggested that, in view of the manifest failure of the Green Revolution, the only hope of the UDCs lay in population control. His statement, you will recall, created an uproar in the underdeveloped countries. Newspaper editorials accused the United States of wishing to prevent small countries from becoming large nations and thus threatening

American hegemony. Politicians asserted that President Kennedy was a "creature of the giant drug combine" that wished to shove its pills down every woman's throat.

Among Americans, religious opposition to population control was very slight. Industry in general also backed the idea. Increasing poverty in the UDCs was both destroying markets and threatening supplies of raw materials. The seriousness of the raw material situation had been brought home during the Congressional Hard Resources hearings in 1971. The exposure of the ignorance of the cornucopian economists had been quite a spectacle—a spectacle brought into virtually every American's home in living color. Few would forget the distinguished geologist from the University of California who suggested that economists be legally required to learn at least the most elementary facts of geology. Fewer still would forget that an equally distinguished Harvard economist added that they might be required to learn some economics, too. The overall message was clear: America's resource situation was bad and bound to get worse. The hearings had led to a bill requiring the Departments of State, Interior, and Commerce to set up a joint resource procurement council with the express purpose of "insuring that proper consideration of American resource needs be an integral part of American foreign policy."

Suddenly the United States discovered that it had a national consensus: population control was the only possible salvation of the underdeveloped world. But that same consensus led to heated debate. How could the UDCs be persuaded to limit their populations, and should not the United States lead the way by limiting its own? Members of the intellectual community wanted America to set an example. They pointed out that the United States was in the midst of a new baby boom: her birth rate, well over 20 per thousand per year, and her growth rate of over one per cent per annum were among the very highest of the developed countries. They detailed the deterioration of the American physical and psychic environments, the growing health threats, the impending food shortages, and the insufficiency of funds for desperately needed public works. They contended that the nation was clearly unable or unwilling to properly care for the people it already had. What possible reason could there be, they queried, for adding any more? Besides, who would listen to requests by the United States for population control when that nation did not control her own profligate reproduction?

Those who opposed population controls for the U.S. were equally vociferous. The military-industrial complex, with its all-too-human mixture of ignorance and avarice, still saw strength and prosperity in numbers. Baby food magnates, already worried by the growing nitrate pollution of their products, saw their market disappearing. Steel manufacturers saw a decrease in aggregate demand and slippage for that holy of holies, the Gross National Product. And military men saw, in the growing population-food-environment crisis, a serious threat to their carefully nurtured Cold War. In the end, of course, economic arguments held sway, and the "inalienable right of every American couple to determine the size of its family," a freedom invented for the occasion in the early '70s, was not compromised.

The population control bill, which was passed by Congress early in 1974, was quite a document, nevertheless. On the domestic front, it authorized an increase from 100 to 150 million dollars in funds for "family planning" activities. This was made possible by a general feeling in the country that the growing army on welfare needed family planning. But the gist of the bill was a series of measures designed to impress the need for population control on the UDCs. All American aid to countries with overpopulation problems was required by law to consist in part of population control assistance. In order to receive any assistance each nation was required not only to accept aid, but also to show progress in reducing birth rates. Every five years the status of the aid program for each nation was to be re-evaluated.

The reaction to the announcement of this program dwarfed the response to President Kennedy's speech. A coalition of UDCs attempted to get the U.N. General Assembly to condemn the United States as a "genetic aggressor." Most damaging of all to the American cause was the famous "25 Indians and a dog" speech by Mr. Shankarnarayan, Indian Ambassador to the U.N. Shankarnarayan pointed out that for several decades the United States, with less than six per cent of the people of the world had consumed roughly 50 per cent of the raw materials used every year. He described vividly America's contribution to worldwide environmental deterioration, and he scathingly denounced the miserly record of United States foreign aid as "unworthy of a fourth-rate power, let alone the most powerful nation on earth."

It was the climax of his speech, however, which most historians claim once and for all destroyed the image of the United States. Shankarnarayan informed the assembly that the average American family dog was fed more animal protein per week than the average Indian got in a month. "How do you justify taking fish from protein-starved Peruvians and feeding them to your animals?" he asked. "I contend," he concluded, "that the birth of an

American baby is a greater disaster for the world than that of 25 Indian babies." When the applause had died away, Mr. Sorensen, the American representative, made a speech which said essentially that "other countries look after their own self-interest, too." When the vote came, the United States was condemned.

This condemnation set the tone of U.S.-UDC relations at the time the Russian Thanodrin proposal was made. The proposal seemed to save themselves and humiliate the United States at the same time; and in human affairs, as we all know, biological realities could never interfere with such an opportunity. The scientists were silenced, the politicians said yes, the Thanodrin plants were built, and the results were what any beginning ecology student could have predicted. At first Thanodrin seemed to offer excellent control of many pests. True, there was a rash of human fatalities from improper use of the lethal chemical, but, as Russian technical advisors were prone to note, these were more than compensated for by increased yields. Thanodrin use skyrocketed throughout the underdeveloped world. The Mikoyan design group developed a dependable, cheap agricultural aircraft which the Soviets donated to the effort in large numbers. MIG sprayers became even more common in UDCs than MIG interceptors.

Then the troubles began. Insect strains with cuticles resistant to Thanodrin penetration began to appear. And as streams, rivers, fish culture ponds and onshore waters became rich in Thanodrin, more fisheries began to disappear. Bird populations were decimated. The sequence of events was standard for broadcast use of a synthetic pesticide: great success at first, followed by removal of natural enemies and development of resistance by the pest. Populations of crop-eating insects in areas treated with Thanodrin made steady comebacks and soon became more abundant than ever. Yields plunged, while farmers in their desperation increased the Thanodrin dose and shortened the time between treatments. Death from Thanodrin poisoning became common. The first violent incident occurred in the Canete Valley of Peru, where farmers had suffered a similar chlorinated hydrocarbon disaster in the mid-'50s. A Russian advisor serving as an agricultural pilot was assaulted and killed by a mob of enraged farmers in January, 1978. Trouble spread rapidly during 1978, especially after the word got out that two years earlier Russia herself had banned the use of Thanodrin at home because of its serious effects on ecological systems. Suddenly Russia, and not the United States, was the *bête noir* in the UDCs. "Thanodrin parties" became epidemic, with farmers, in their ignorance, dumping carloads of Thanodrin concentrate into the sea. Russian advisors fled, and four of the Thanodrin plants were leveled to the ground. Destruction of the plants in Rio and Calcutta led to hundreds of thousands of gallons of Thanodrin concentrate being dumped directly into the sea.

Mr. Shankarnarayan again rose to address the U.N., but this time it was Mr. Potemkin, representative of the Soviet Union, who was on the hot seat. Mr. Potemkin heard his nation described as the greatest mass killer of all time as Shankarnarayan predicted at least 30 million deaths from crop failures due to overdependence on Thanodrin. Russia was accused of "chemical aggression," and the General Assembly, after a weak reply by Potemkin, passed a vote of censure.

It was in January, 1979, that huge blooms of a previously unknown variety of diatom were reported off the coast of Peru. The blooms were accompanied by a massive die-off of sea life and of the pathetic remainder of the birds which had once feasted on the anchovies of the area. Almost immediately another huge bloom was reported in the Indian ocean, centering around the Seychelles, and then a third in the South Atlantic off the African coast. Both of these were accompanied by spectacular die-offs of marine animals. Even more ominous were growing reports of fish and bird kills at oceanic points where there were no spectacular blooms. Biologists were soon able to explain the phenomena: the diatom had evolved an enzyme which broke down Thanodrin; that enzyme also produced a breakdown product which interfered with the transmission of nerve impulses, and was therefore lethal to animals. Unfortunately, the biologists could suggest no way of repressing the poisonous diatom bloom in time. By September, 1979, all important animal life in the sea was extinct. Large areas of coastline had to be evacuated, as windrows of dead fish created a monumental stench.

But stench was the least of man's problems. Japan and China were faced with almost instant starvation from a total loss of the seafood on which they were so dependent. Both blamed Russia for their situation and demanded immediate mass shipments of food. Russia had none to send. On October 13, Chinese armies attacked Russia on a broad front. . . .

A pretty grim scenario. Unfortunately, we're a long way into it already. Everything mentioned as happening before 1970 has actually occurred; much of the rest is based on projections of trends already appearing. Evidence that pesticides have long-term lethal effects on human beings has started to accumulate, and recently Robert Finch, Secretary of the Department of Health, Education and Welfare expressed his extreme apprehension about the pesticide situation. Simultaneously the petrochemical industry continues its unconscionable poison-peddling. For instance, Shell Chemical has been carrying on a high-pressure campaign to sell the insecticide Azodrin to farmers as a killer of cotton pests. They continue their program even though they know that Azodrin is not only ineffective, but often *increases* the pest density. They've covered themselves nicely in an advertisement which states, "Even if an overpowering migration [sic] develops, the flexibility of Azodrin lets you regain control fast. Just increase the dosage according to label recommendations." It's a great game—get people to apply the poison and kill the natural enemies of the pests. Then blame the increased pests on "migration" and sell even more pesticide!

Right now fisheries are being wiped out by over-exploitation, made easy by modern electronic equipment. The companies producing the equipment know this. They even boast in advertising that only their equipment will keep fishermen in business until the final kill. Profits must obviously be maximized in the short run. Indeed, Western society is in the process of completing the rape and murder of the planet for economic gain. And, sadly, most of the rest of the world is eager for the opportunity to emulate our behavior. But the underdeveloped peoples will be denied that opportunity—the days of plunder are drawing inexorably to a close.

Most of the people who are going to die in the greatest cataclysm in the history of man have already been born. More than three and a half billion people already populate our moribund globe, and about half of them are hungry. Some 10 to 20 million will starve to death *this year*. In spite of this, the population of the earth will increase by 70 million souls in 1969. For mankind has artificially lowered the death rate of the human population, while in general birth rates have remained high. With the input side of the population system in high gear and the output side slowed down, our fragile planet has filled with people at an incredible rate. It took several million years for the population to reach a total of two billion people in 1930, while a *second two billion will have been added by 1975!* By that time some experts feel that food shortages will have escalated the present level of world hunger and starvation into famines of unbelievable proportions. Other experts, more optimistic, think the ultimate food-population collision will not occur until the decade of the 1980's. Of course more massive famine may be avoided if other events cause a prior rise in the human death rate.

Both worldwide plague and thermonuclear war are made more probable as population growth continues. These, along with famine, make up the trio of potential "death rate solutions" to the population problem—solutions in which the birth rate-death rate imbalance is redressed by a rise in the death rate rather than by a lowering of the birth rate. Make no mistake about it, *the imbalance will be redressed.* The shape of the population growth curve is one familiar to the biologist. It is the outbreak part of an outbreak-crash sequence. A population grows rapidly in the presence of abundant resources, finally runs out of food or some other necessity, and crashes to a low level or extinction. Man is not only running out of food, he is also destroying the life support systems of the Spaceship Earth. The situation was recently summarized very succinctly: "It is the top of the ninth inning. Man, always a threat at the plate, has been hitting Nature hard. It is important to remember, however, that Nature bats last."

The Web of Life

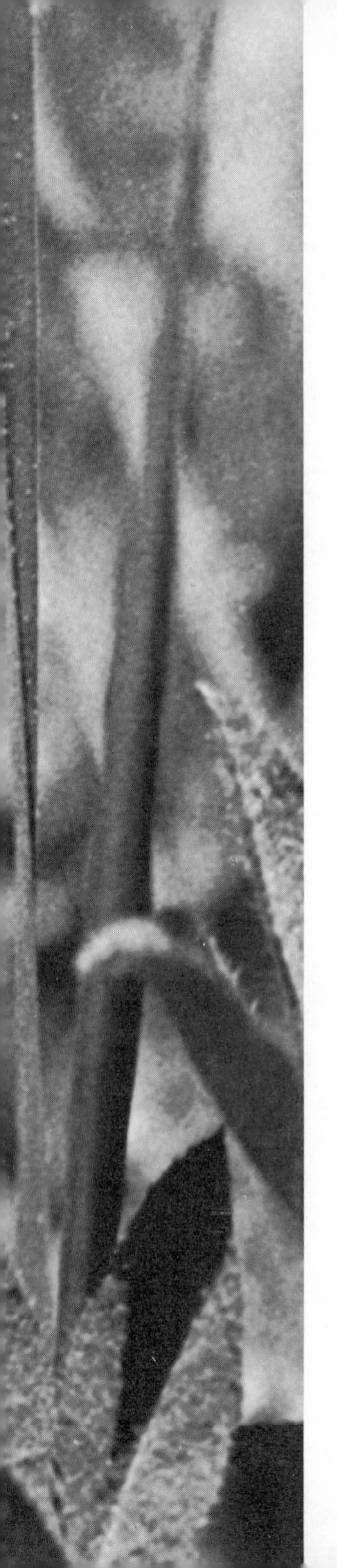

Ecology is the science which studies the relationships between organisms and their environment. Seizing upon this definition, U. S. Senator Ted Stevens of Alaska derided the efforts of conservationists who had come to Alaska recently to examine the effects of the oil boom on the northern coast of that state. "There are no living organisms on the North Slope," stated the Senator.

The Senator's fundamental ignorance that a web of life exists even in as inhospitable a region as the North Slope of Alaska is a reflection of how unaware he was of the facts of his own environment. Barry Weisberg points out in his article on oil in Alaska (page 187) the possible environmental effects of a technological invasion upon the North Slope. In the following essay, Robert and Leona Rienow provide similar examples of ecological systems (ecosystems) which have been ravaged by man acting without understanding the consequences of his actions. The Rienow essay is a good starting place for the layman interested in ecology because of its wealth of specific information and its many examples drawn from both history and the present.

This essay is a chapter from the Rienows' book Moment in the Sun, *a study of the declining quality of our environment. Robert Rienow (1909-) is a professor of political science who has long been associated with conservation movements. His wife, Leona, has collaborated with him on this and other books.*

ECOLOGY: THE RELENTLESS SCIENCE

by Robert Rienow and Leona Train Rienow

Basic to an understanding of what a crowded and crumbling environment can do to man is an understanding of ecology, the science of "the mutual relationship between organisms and their environment." It is the science of man's relationship to all living things and to the very earth itself.

Supreme Court Justice Abe Fortas, in accepting the Albert Schweitzer medal from the Animal Welfare Institute in November, 1965, had this to say: "Life is a seamless web. It connects us not merely with one another, but with all that is sentient—with all that shares its miracle of birth and feeling and death." And Fairfield Osborn, in his small masterpiece *Our Plundered Planet,* declared as a "flat statement" that "in a world devoid of other living creatures, man himself would die."

Thoreau so identified himself with his beloved fields, woods, and lake that he became an integral part of them in mood, in pulse, in sensation. Almost every great or thoughtful mind has involved itself deeply with nature and earthy things, with all other forms of life.[1] It is well known that both Presi-

---

dents Washington and Jefferson loved their acres and the natural beauty that grew on them to the point of obsession.

But it is not so well known how closely Lincoln associated himself with nature, how attuned he was to other forms of life. Historian Coyle in *The Ordeal of the Presidency* tells a moving story about this unfathomable man.

> *After the fall of Richmond, a young French diplomat, the Marquis de Chambrun, was riding through Petersburg in a carriage with Lincoln, when he stopped the carriage to look at a noble white oak with gnarled wide-reaching arms. Chambrun was deeply impressed by Lincoln's feeling for the tree and recorded that "he talked as if he might be some kind of a tree himself."*[2]

Thoreau's most famous quotation is "In wildness is the preservation of the world," and he believed that artificiality brings decay of virility and then death. True or not, it is an indisputable and proven fact that we cannot beat up our environment without beating up ourselves. This law is relentless. Though the consequences may be delayed, be forestalled, even masked, they will always strike and often to kill.

In the old days when the lovely Schoharie Valley in upstate New York was a garden of hops, skunks were honored. It was said that natives "would shoot anybody who shot a skunk." The reason was both ecological and economic: the skunk was, in nature's plan, the enemy of the hop grubs and in that role the defender of the Valley's enterprise.

The insight of these settlers is rare; men and civilizations, past and present, have been blind to the proper place of fellow creatures and lesser life forms in the pattern of existence. From the Head Office of the Royal Bank of Canada in May of 1960 came a monthly newsletter profoundly concerned about our neglect of our environment. Said the letter:

> *The forces set in motion by every act of every animal and bacterium, by every inch added to the growth of plant or tree, affect the lives of other creatures. The principles which govern these interrelationships are embraced in the science called ecology, a word coming from the Greek for "home" or "estate." Ecology is the study of how the household of nature is kept in order.*

Why are the salmon gone from Lake Ontario and the bison from the plains, the deep three-foot layers of humus washed to the sea, and the climate altered? Professor A. F. Coventry, speaking to the Toronto Field Naturalists' Club, warns, "We have for a long time been breaking the little laws, and the big laws are beginning to catch up with us."

Great civilizations have dwindled to mediocrity or collapsed like a rotten gourd for violation of the big laws of ecology.[3] Professor Raymond Bouillenne, director of the Botanical Institute, University of Liège, who knows the African scene intimately, traces this sorrowful trend.[4] Where are Cyrenaica's gardens of Berenice, once so famed in Rome? The Libyan Desert hides the ruins of great cities such as Thyṣdrus, whose sport stadium once held 60,000 men. Beneath thick layers of Sahara sand the French explorer August Chevalier found traces of dense forests that existed less than 2,000 years ago. What happened to the fabled glory of Arabia, Babylon, and Tibet? In Morocco alone, since the Roman period, "12 ½ million acres of forests have disappeared as a result of fire and overgrazing by sheep and goats."

Egypt of the golden tombs, whose rich fields once spread far over the land, now shrinks and starves within the valley of the Nile. China, once lush with forests and the fruits of vine and twig, has watched her deserts march in, driving her too fecund people before them into the last refuges of the valleys of the Blue and Yellow Rivers.

Was it only a million years ago that the first man stood upright? Was it only eight thousand years ago that man began to till? In the now classic volume *This Is the American Earth,* which Nancy Newhall scripted and Ansel Adams photographed, fact after shocking fact hits the reader in the face as it traces the history of man's destruction of Earth down through the millennia.

It was thousands of years ago that men devastated, with the help of uncontrolled seas of sheep, the rich, lake-dotted lands of northern Mongolia—now despised wasteland whose very name repels. Can we believe that Greece was once an Eden of forests and waters; that Palestine was brought from overflowing richness to famine, drought, pestilence; that in Europe wilderness was wiped out by men before the fifteenth century?[5]

When the avid human tide swept over North America it laid waste not only to the level forest lands but reached up the mountains as well. "Thank God," cried Thoreau, "they cannot cut down the clouds!" Most of the mammoth kills of bighorn sheep, buffalo, elk, caribou, moose, bears, and antelope were for boasting; the pioneers massacred buffalo to take out the tongue and left the rest to rot. In the Midwest dynamite thrown into the thousands of teeming lakes destroyed forever the ecological colonies that had been building up since the recession of the last glacier.

But all ecological sins come home to roost, and bitterly indeed have the once richly endowed civilizations of the past paid for their rape of nature. In America the long overdue account of our disregard for ecological laws is only now beginning to be presented. The payments must be made, if not in cash, then in blood.

The essence of technological, if not civilized, pursuits is the constant clashing with ecological principle. A wheat field or an apple orchard is, by itself, an invitation to ecological breakdown. They offer no defenses in depth to the attack of pests or diseases.

On the other hand, Professor Samuel Graham cites the protective qualities of naturally mixed stands of trees, for instance, "In the North, where you have the familiar mixture of beech, birch, maple, hemlock, and pine, you seldom find damage done to the hemlock. But where the species exists in pure stands it may be destroyed in a single season by the hemlock looper."[6] The pest's commuting distance is shortend. He has been spread a feast, and joyfully he multiplies. When, as in New Brunswick, a spray program for control of the spruce budworm is launched on the headwaters of the Miramichi River, no thought is given to the river's salmon; shock follows when we find the river clotted with the bodies of 90 per cent of the one- and two-year-old fish.

We let the great cats in our western forests claw open the landscape on the upper reaches of waterways and are stunned when nature reacts with a violent and choked flood. At Christmastime of 1964 the west coast of California, where unrestricted redwood and Douglas fir logging is carried on, was the tragic scene of very damaging floods.

But in the very heart of the area a miracle took place. It happens that Nature Conservancy owns and protects in its natural state the whole of the watershed of Elder Creek in Humboldt County. The report on the Northern California Coast Range Preserve said: "Elder Creek, as an uncut watershed, ran clear even though extremely high. Observers from airplanes reported that it stood out as an exception among the roiled waters of the other streams in the region."

Still these destructive trends continue unabated over the entire world. Bouillenne terms the regression of the forest massif of Central Africa "disastrous." He notes that in the Congo, where the cover is just too delicate to withstand clearing for hunting and cultivation, 30,000 square kilometers of soil have been destroyed in six years.

> *In Madagascar, the drama has been played out over a period of 60 years. This island was once covered with splendid forests; today 70 per cent of its area is occupied by an ocean of tough grasses, ravaged by fire and unsuitable even for the feeding of herds. In short, we are in the throes of an apparently irreversible progressive* reduction *of the surface of cultivable lands.*[7]

What happened to the great oak forests of the Hittite Empire, where the conqueror Tamerlane hid his herd of elephants? Where are the groves of cypress and palms that once made lush the desolate plains of Iraq? Why is there no trace of the vanished forests of Phoenicia, whose stout timbers long ago raised this people to greatness as a naval power? When the 35,000 men of Tyre cut Lebanon cedar for the Temple of Solomon, how is it that this thriving tree community never reseeded and recovered from the shock? And why has the Sahara Desert been moving southward—for the last 500 years—at the rate of at least a mile a year on a wide front of some 1,800 miles (the estimate of K. H. Oedekoven, forestry researcher for the U.N. Food and Agriculture Organization)?

The answer is, of course, that when man scalps the land he outrages the ecology; he destroys the humus, evaporates the water, erodes the soil, alters the climate, and so shocks the environment that never again can the conditions that brought these choice climax forests into being prevail.

Here, then, is a brief sampling of ecological relationships. Put in the words of Phillip Keller the lesson is: "We make our greatest mistake when we believe that the world belongs to us. It does not—we belong to it!"

Much of our small sinning against the laws of ecology is through ignorance. We have an inbred passion for killing snakes—even the harmless and ecologically precious black snake, which can slither its way into rodents' holes and attack them in their very retreats. California farmers, understanding that they would be overrun with rodents were it not for this valuable ally, accost the museum collectors out gathering snake specimens with the stern bag limit of one snake per farm.

Then there is the common bluejay, whom most people resent because of the manner in which it gulps whole sunflower seeds by the pound. But the bluejay pays back society in a unique way; foresters have discovered it is nature's most ardent "reforester." The jay not only places the seed of trees in the ground with great care, but it also covers the spot with leaves and pine needles as a protection from the weather. Taking a cue from the bluejay, at least one state of the Union supplies each hunting license buyer with a package of pine seeds to be scattered in waste areas on his hunting expeditions, an idea that Theodore Roosevelt would have considered "bully, indeed!" And although Mother Nature invented poison berries and seeds, she created sixty-three kinds of birds with consuming appetite for the baneful things, which foresight probably saved humanity in the form of *Pithecanthropus erectus* from an early extinction!

Often the account of the interrelation of living forms is steeped in drama. Thus it happened that one year the citizens of Cattaraugus, Erie, Wyoming, and Allegheny Counties in New York State were witnessing with horror an army of inchworms avidly devouring their forests. Ten thousand more acres were in the path of the marching worms when without bugle calls and seemingly from nowhere a counterforce of birds, field mice, shrews, and the Calasona beetle was suddenly mustered. Within days the infestation was over.

The intricacies of the world we live in are underlined by the report of three Cornell researchers—Thomas Eisner, Rosalind Alsop, and George Ettershank—who studied the webs of orb-weaving spiders. Designed as a special trap for flying insects, these webs have a framework of ordinary threads interlaced with adhesive strands of a very sticky quality—a device that should have doomed the unwary insect without hope of disentanglement.

But what did the researchers find? That some moths have been provided with detachable scales on their wings and bodies. Merely by leaving a few excess scales behind on a sticky strand they can thwart the hungry expectant spider. Other moths are covered with a powder with which they can coat the sticky drops and so make their Houdini-like escape.

Thus not only is the minor world of insects and spiders complex, but it maintains a fair balance. Spiders have, over the millennia, trapped insects. "But over the many millions of years insects have been preyed upon by spiders, the insects have developed equally 'ingenious' evasions."[8]

This intricacy of relationships in nature bestows benefits of which man is too often totally ignorant. Speaking of a hated and harassed bird, the Red Tail hawk, ornithologist Alexander Sprunt, Jr., says this:

> *The food of the Red Tail is mainly small mammals and not poultry as it is often believed. Birds constitute no more than 10*

per cent of its food, while an examination of 562 stomachs showed remains of mammals (principally meadow mice) in 409. Reptiles, insects, and crawfish are also taken.[9]

During the breeding season a single Red Tail will bring 1,000 mice to its nest to feed its ravenous young. Without its appetite and that of other hawks and birds of prey the world would be ankle-deep in rodents, and women would have to live on top of stepladders.

But perhaps one of the finest examples of the competence of nature and the bungling of man is the story of southern Oregon's pine country, where there has been a serious population explosion of porcupines. The porcupine, which delights in girdling a young pine just as the tree is entering its best growing stage, has a place in nature and undoubtedly even a right to its share of pine trees. But it was not intended that porcupines should preempt the Northwest any more than that man should monopolize the globe. Why, then, was this happening?

Then old foresters recalled that once the porcupine had a fierce and insatiable enemy, a sleek big cousin of the mink, who had roamed these forests with a special purpose of harassing the prickly rodent. This was the fisher, who, of all animals who might have a taste for juicy fat porky flesh, had also the rare ability to leap from one tree top to another, stalk his quarry, and dispatch it without being needled painfully in the process. It was a very special art. But if the fisher relished porky, man relished the fisher's silky fur even more. Thoughtlessly, he trapped the species out of the forests, and then wailed to watch his forests destroyed.

It is not always so direct a disturbance of nature's plan that causes trouble. Most often man upsets the habitat to which a creature is highly attuned. Consider the giant California condor—the "Thunderbird" of Indian lore—whose dark bulk once soared lazily over many parts of the Western sky. The condor must have cliffs for nests, where he is provided with just the right air currents to bear his heavy body aloft in his search for carrion. Brooking no violation of his privacy, he will abandon a nest should human traffic come within earshot.

Because he demands a tiny but inviolate corner of this world for himself, the condor seems doomed. Poisoned and hunted elsewhere, the forty remaining birds have retreated to the Sespe Wildlife Refuge of 53,000 acres, which man reluctantly assigned them in 1951. Even here he is being threatened. There are those who think we need another impoundment of water on the Sespe River more than we need the condor. In this case we cannot claim that we did not know the ecology of the species; we are simply—and callously—ignoring it.[10]

We can put in our dam and create still another undistinguished development and drive the condor to extinction. For some unfathomable reason we think this makes us richer than the Indian who stood in silent awe as the Thunderbird soared far overhead, then floated, with its weird and Harpy grace, into the canyon's swirling mists.

Ecosystems need not be as big or as dramatic as that of the fisher-porcupine-pine tree relationship. They need not cover the thousands of acres of a refuge edged with misted cliffs. The drama of the life chain can be excitingly enacted within the environs of a dead tree. In the design of nature, where nothing is ever lost or wasted, it is imperative that the tree, which removed chemicals and other materials from the soil, return them so that they can be used over gain.

Thus, in the little world of a dead tree, shelf fungi first spring up to speed the decomposition with the enzymes they secrete. Then they, in turn, provide wastes that green plants can feed upon. Now insects join in the attack; carpenter ants and wood-boring beetles invade, channeling out vast cities of tunnels, and opening thousands of doors to the weather and to bacteria. Then woodpeckers in pursuit of the insects industriously peck out larger and larger cavities.

Ultimately the rotting tree falls to the ground, creating in its shelter a home and food supply for a whole new host of creatures. In move the armies of slugs, mice, and snails, squirming larvae, worms, and hibernating beetles, to chaw and dig and speed the breakdown of the log. Soon come their camp followers, toads, frogs, snakes, and moles, who sidle in to live happily on this woodland smorgasbord. Even the beautiful mourning cloak butterfly comes there to hibernate—a delicious morsel. The fallen log is palpitatingly alive as it crumbles into humus. In this wondrous, complicated way the tree's "life-giving elements are returned to the earth that gave it birth. A chain of life is ready to begin again."[11]

A scientific team of botany, biology, and zoology professors sought to discover why the giant cactus of the Sonoran Desert, the saguaro, is in trouble. Why, despite its 200-odd fruit, each harboring 200,000 seeds, is this handsome plant failing to hold its own? Again, man was found in the picture. He brought in cattle in the 1880's in such numbers that the low plants, which provide cover for the small saguaro, were eaten away. With sparse plant cover erosion followed, channels were cut, and the scarce water supply drained away.

All this is a common enough ecological chain. But man added new complications for the majestic saguaro. Man hates the coyote, whose regular food consists of desert rodents, notably the ground squirrel, wood rat, and rabbit. The ground squirrel and wood rat feed on the saguaro as much for water as for food. Since, however, it became official government policy to exterminate the coyote by all means (including the frightful 1080 poison), the rodent population has multiplied astronomically. There are now fears that the Saguaro National Monument may not survive.[12] We will have exchanged an eternal wonder for a beefsteak; we will have lost a nesting site for desert birds such as the woodpeckers and elf owls; we will have fostered death and cheated life; we will have taken another edge off the quality of man's experience.

Centuries ago Leonardo da Vinci, in a combination of science and intuition, mused: "In nature is the answer to everything." Alan Devoe, over a decade ago, explained: "All creatures are in a common brotherhood . . . interconnected with everything else. Not only is there a basic brotherhood between [all men] but there is a bond between a man and a mouse, or a tree and a fox, or a frog and a raccoon." And he added: "We are one small ingredient in a whole of unimaginable Vastness . . . a part of a general and embracing interdependence. . . . We are supported by starfish. An owl props us. Earthworms minister to hold us upright."[13]

As early as 1939 Hamilton Basso in *Days Before Lent* put a speech in the mouth of old Dr. Gomez, a Central American revolutionist exiled from his country:

> *Do they not understand that as man subdues nature he subdues himself—that man, being an animal, is as dependent upon the operation of natural laws as an amoeba or a frog? . . . Let the balance necessary to man's existence be destroyed, and it is quite possible that he will go the way of the wild horse and the white-tailed gnu.*[14]

The many years of senseless attack on the coyote with the indiscriminate poison soduim fluoracetate (commonly known as 1080, already mentioned) set up a terrifying sequence of tragedy, The initial target kill is gruesome enough: one ounce mixed with 1,500 pounds of horsemeat makes a lethal bait even if only two ounces of it are consumed by a coyote. An investigative lick may be enough.

The evil of the bait runs through a whole chain of innocent victims. A mouse samples a grain and dies. The fox or bobcat that eats the mouse retches and dies. A bird that consumes the grain droppings will also expire. The poisonous vomit falling upon the grass sickens and kills the animals who graze upon it. And, of course, the buzzards, eagles, crows, and jays or any carnivorous animals that eat the poisoned carrion are doomed. There is a kernel of irony in the fact that a dove, quail, duck, or pheasant that may have consumed the man-placed bait 1080 just before being shot may become a deadly repast on the table of man himself.

Thus, we see that we have ecological "chains of death" as well as "chains of life." A shattering thesis has just been verified by three professors of chemistry, reported at the 1965 meeting of the American Chemical Society at Atlantic City, New Jersey, on September 16. The professors, Donald E. H. Frear, Ralph O. Mumma, and William B. Wheeler, of Pennsylvania State University, and the fact they reported—stoutly ridiculed by scientists and government officials alike for many years—is that forage crops can and do take up insecticides from the soil *through their root systems* and pass it into grazing cattle, thence into man.

The Penn State report, says the Pennsylvania *Game News,* December, 1965, showed that even though a crop *has not been sprayed* in the year of harvest, pesticide residues that were still present in the soil from preceding years were absorbed in the new and quite unsprayed crops through the plants' roots. The scientists grew a number of forage crops in sand and soil permeated with "radioactively tagged" dieldrin, DDT, and other insecticides. Then they watched each insecticide through radioautography as it rose into and distributed itself through the plants!

Can the familiar story of Mao Tse Tung and the sparrows be true? Some years ago the Chinese dictator was led to believe that the sparrows were eating too much of China's precious grain. On his order, so the fantastic tale runs, "millions of Chinese, young, old, and crippled, took poles, bamboo, and brooms and waved them at the sparrows. The frightened birds fluttered into the air and kept flying until they dropped to earth dead of exhaustion."

Whether the birds were frightened to death by a seemingly endless sea of humanity gone suddenly insane, or whether the Chinese actually beat the birds to death, the ecological point was made. The insects that the sparrows had been feasting upon increased so disastrously that the fruit crop was almost ruined. The battle of the sparrows was called off.

In somewhat like manner fishermen in northwest Missouri, ignoring the thrill that the annual migrations of the fresh-water white pelicans gave the observers, and resentful of having to share fish with them, organized raids on the birds' island nests and decimated the population. Today the white pelicans are on the critical list for survival.

But so are the game fish, unfortunately. It so happens that the diet of the pelican is largely rough fish; with the pelican all but wiped out, the rough fish are increasing and crowding out the game fish, ruining the fishermen's sport.

Sir Francis Bacon once said, "To learn to dominate Nature we must first learn to obey her." And to understand her laws, it might be added. "Men are running wild through the biological world like some malignant agent," warns Dr. Marston Bates, the renowned naturalist of the University of Michigan. "Everything man touches turns to garbage." Or to corpses.

The Committee on National Resources of the National Academy of Sciences puts man's role in more dignified but no less condemning terms:

> *Man is altering the balance of a relatively stable system by his pollution of the atmosphere with smoke . . . , alteration of the energy and water balance . . . , over-grazing, reduction of evapotranspiration, irrigation, drainage . . . , building of cities and highways; by his clearing forests and alterations of plant surface cover, changing the reflectivity of the earth's surface and soil structures; by his land-filling, construction of buildings and seawalls, and pollution, bringing about radical changes in the ecology of estuarine areas; by the changes he effects in the biological balance . . . , the erection of dams and channel works; and by the increasing quantities of carbon dioxide an industrial society released to the atmosphere.*

It is hard to tell whose indictment is the harshest. Aldous Huxley, famed English author, has accused: "Committing that sin of overweening bumptiousness, which the Greeks called *hubris,* we behave as though we were not members of the earth's ecological community, as though we were privileged and in some sort, supernatural beings and could throw our weight around like gods." All of which seems to substantiate Leonardo da

Vinci's no-nonsense damning of glorious *Homo sapiens:* "Perfidious Man! Cursed is the day you left the cave."

Undoubtedly Da Vinci had had a bad morning; perhaps somebody had sat down on his palette. It was inevitable that man would alter his environment to increase his convenience. It is less excusable that he has been both so callous and cruel in his alterations that all life must now pay dearly.

Perhaps if man nurtured greater understanding of the role of each of his fellow creatures on earth he might act more cautiously in the destruction of any life. When he picks the despised stinkbug off the berry cane to crush, does he realize its value as a repellent of lice and other pests? When he shoots a brown thrasher in his garden does he know that this bird accounts for nearly 6,000 bugs a day? What kind of double talk is it, asks nature writer Ferris Weddle, to exult in the killing of a cougar because "it saves a dozen deer" and in the next breath to bewail the "over-population" of deer causing starvation from lack of browse? Do California taxpayers (whom, says Fish and Game Director Walt Shannon, it costs $629 per lion in costs and bounties for the killing campaign) still see no inconsistency here?[15]

One hundred and twenty years ago Michigan placed a bounty on timber wolves and all but exterminated them; there are only about forty left in the entire state. But twenty of these live safely on Isle Royale. As a consequence, the moose on Isle Royale, far from being diminished by the wolves, are developing into a beautiful superbreed. The wolves have served nature's purpose of eliminating the runts, the old and diseased.

In a less direct fashion we have the now classic story of the boy who set out to trap the skunks in the vicinity of his duck pond. After awhile his ducks began to vanish mysteriously, one by one. It took an old woodsman who understood the web of life to explain to the boy that skunks eat snapping turtle eggs, and that when the skunks are killed off the turtles increase and quietly pick off the ducks. Akin to this ecological ring is the grasshopper-prairie dog chain, described recently by zoology Professor George M. Sutton of Oklahoma University. Burrowing owls feast on grasshoppers and similar pests. But they use the burrows of prairie dogs for their nests. When ranchers, seeing the prairie dogs nibbling at their pastures, persuaded federal agents to start a full-scale chemical warfare against the prairie dogs, the burrowing owl population decreased. Soon armies of grasshoppers and other insects began to take over and the end is not yet come. It will be interesting to follow what happens.

But ecological thinking is new, and there are few indeed who take any heed of end results when they decide to tamper with their environment. "The utilitarians," notes naturalist and Professor Dan McKinley of Lake Erie College, Ohio, "aim [at] as many people as possible, all carefully regulated as to hours of labor, leisure, and lechery. As they quibble, unique organisms . . . such as bogs, deserts, and islands are carelessly lost and haphazardly succeeded by the stability of ragweeds, starlings, barn rats, and eroded soils."

Were it not for predators such as owls this continent would be covered with two and a half inches of mice from coast to coast within a year, say the statisticians. Senator Gaylor Nelson tells of a bird watchers' paradise—a little island covered with sooty terns. Someone released a few rats on that island which multiplied and cleaned up all the terns without much ado. Rats are almost as destructive as men.

Botanist Kenneth W. King of Antioch College recommends that we adopt Rachel Carson's "biocentric concept" for our own well-being. And Aldo Leopold long ago pleaded for the "ecological conscience" that would place man in a less arrogant but truer spot in the Creator's scheme of things. Charles A. Lindbergh came to grips with his relationship to his environment in the jungles of Africa, far from the distractions of civilization. "Lying under an acacia tree with the sounds of the dawn around . . . I became more aware of the basic miracle of life," he wrote. "Not life as applied humanly to man alone, but life as diversified by God on earth with superhuman wisdom—forms evolved by several million centuries of selection and environment. . . . I realized that if I had to choose, I would rather have birds than airplanes. . . ."[16]

But perhaps ecological awareness is growing at last. The high party official in China who instigated the bird-extermination idea has been dispatched to become foreman of a land-clearing project in remote, icy, seldom-come-back-from, northwestern China. And in Russia a strange edict has been issued. Dolphins, it seems, enjoy a very elaborate sound communication system, which some scientists call "delphinese." (According to a news item: "It is of interest to note that while some dolphins are reported to have learned up to fifty words in English—used in correct context—no human has been reported to have learned delphinese.") The Soviet Government announced in March of 1966 that henceforth the catching and killing of dolphins was banned because "their brains are strikingly close to our own."

In a more serious vein we are now told that 1080 has at last been discredited in the West. And in the prairie states, notably North Dakota, natives are beginning to understand the ecology of trees and soil from the great shelterbelt planting of the 1930's. Luxuriating now in "beautiful lanes of verdure, holding down the soil, conserving snow moisture, reducing fuel bills, and giving beauty to the landscape," they are again planting 11,000 acres a year in seventy-three soil conservation districts, writes State University Extension Service forester John Zaylskie.

Former Governor Karl F. Rolvaag of Minnesota, a leader among governors, rescinded the bounty on timber wolves.[17] On the side of industry Parke, Davis (but for one example) collects soil samples and isolates 34,000 cultures of microorganisms a year in the hope of making the exciting discovery of a new cure for some still unconquered disease. And one of the many other endeavors is that of scientist A. W. Kuchler, who is leading a vast mapping of the nation's vegetation to give us insight into the patterns of existence, a needed inventory of our heritage of life forms and their relationship to man.[18]

"Less than a century divides the era when America was looked upon as a Garden of Eden or savage wilderness and the time when it took first place as the world's industrial giant," historian Arthur A. Ekirch, Jr., reminds us. "Probably no people have ever so quickly subdued their natural environment. . . ."[19] They have, moreover, subdued it with a brutality and lack of foresight that have left the world gasping. Yet most scholars still pretend an abiding faith in our American future.

However, it is almost without exception a faith in man working with nature, not in his ability to continue the pitiless and pitiful fight against her. The great engineer-designer Richard Buckminster Fuller, who launched the International Union of Architects' ten-year plan of "world redesign," says: "The universe is so successful, I simply want to learn its principles and apply them rather than to exploit it blindly and fear for survival."

"Man," says natural scientist F. Raymond Fosberg, director of Nature Conservancy, "is doomed to extinction if he cannot be persuaded of his dependence on an intricate life process and his need to respect and protect that process at every step in order to deserve its respect and protection." Where only a few years ago popular naturalists like Bernard de Voto and "Ding" Darling stood almost alone, ecologists are today buttressed by the biggest brains in publishing, industry, and government. Department of the Interior's old and hated Bureau of Predator and Rodent Control has been rebaptized the Division of Wildlife Service, and its approach greatly broadened in the public interest.

We do not live to ourselves; man was not meant to live to himself.

When Dr. Alfred G. Etter, western representative of Defenders of Wildlife, lost his dog to strychnine set out by a State Fish and Game agent "to kill a few magpies to save a few pears," he wrote a touching memorial. Some of it went as follows:

> *It is strange how unimportant each of us may seem in the great fairy ring of life, yet how good it feels to touch a dog, to hug her big head, to feel the beating tail against your leg. In the great range of time and earth, it is good to know the feel of other life, to wipe away the loneliness of being man.*
>
> *'For that which befalleth the sons of man befalleth the beast; even one thing befalleth them: as the one dieth, so dieth the other; yea, they have all one breath; so that a man hath no preeminence above a beast; for all is vanity.'*
>
> *Eccles. 3:19*

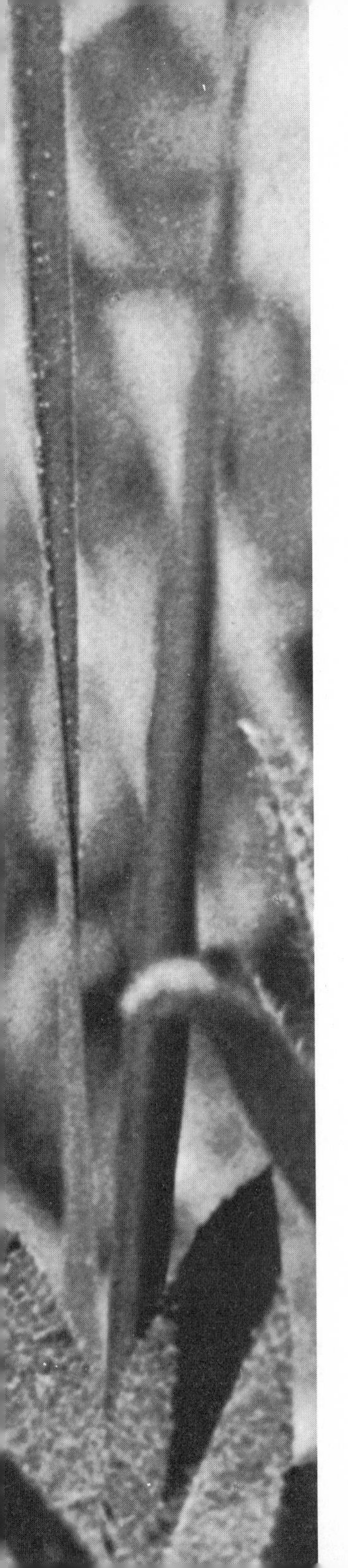

Unlike the Rienows, Marston Bates (1906-) is a professional biologist, a professor of zoology. His approach to ecology in this essay contrasts to some degree with the highly specific, anecdotal information presented by the Rienows. Bates is more often theoretical and philosophical, searching for means by which the entire human ecosystem can be understood through a greater awareness of its many interconnecting relationships. He defines and examines the concept of the ecosystem. He cites the dangers which may result when man shortens and simplifies ecosystems to suit himself. He points out that complex communities are more stable than simple ones, and that the human tendency to simplify ecosystems can lead to the kinds of eco-disasters which have occurred in such closed ecosystems as small islands. For, as Bates shows, the earth itself is a closed ecosystem, an island or a "spaceship." The metaphor of the earth as spaceship, a closed, intradependent system hurtling through space, is one that will be seen frequently in this book, and one that is explored specifically by Kenneth Boulding in "The Economics of the Coming Spaceship Earth" (page 307).

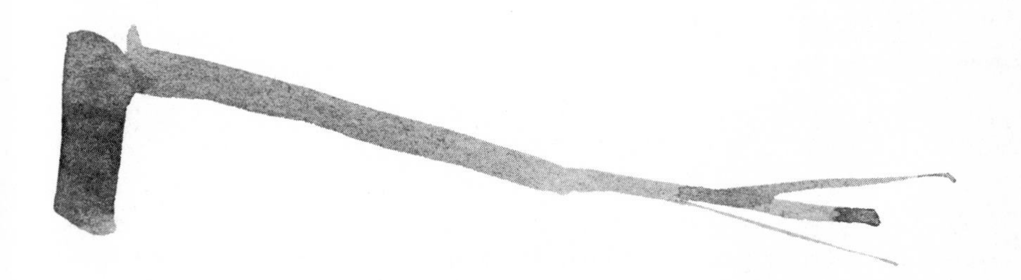

THE HUMAN ECOSYSTEM

by Marston Bates

"In the West, our desire to conquer nature often means simply that we diminish the probability of small inconveniences at the cost of increasing the probability of very large disasters." [1]

Our planet has been aptly called "Spaceship Earth." It forms, overwhelmingly, a closed system as far as materials are concerned. Science fiction to the contrary, we have no present basis for believing that this essential isolation will be altered—that we can colonize other parts of the solar system or import from outer worlds any appreciable quantities of materials. This earth is our habitat and probably will be as long as our species survives. We would do well, then, to treat it carefully and to take thought in planning our actions.

Early men—the food-gathering and hunting peoples of the Old Stone Age—were closely interacting parts of the biological communities in which

---

they lived. Tools and language made them unusually efficient hunters but not really different in their impact on the community from other kinds of social carnivores or omnivores. The first major change in man's relations with nature came with the deliberate making of fire. No other animal starts fires. Their frequency must have increased greatly when man began setting them, with far-reaching ecological effects.

Man's relations with the rest of nature underwent a much greater change with what anthropologists call the "Neolithic Revolution"—although "revolution" is perhaps a poor word for changes that may have taken millenia and that occurred at different times and in different sequences in various parts of the world. The important changes, which were at least in part interrelated, involved the cultivation of plants, the domestication of animals, the settlement of villages, the making of pottery, and a series of improvements in toolmaking. Man began to alter the biological community in which he lived by removing vegetation he did not want or could not use and replacing it with crops.

Gordon Childe,[2] the British anthropologist, looked at post-Neolithic developments in terms of two further "revolutions"—the Urban, turning on the transport and storage of food, which made possible the formation of cities and the specialization of labor and knowledge, and the Industrial, based on the harnessing of power other than the muscles of men or beasts. C. P. Snow[3] would add a recent revolution, the Scientific, resulting from the union of science and technology for the solution of practical problems —the revolution that gave rise to the dramatic and drastic increase in power available to man in the twentieth century, including the development of nuclear and electronic techniques. The Scientific Revolution in particular has the potential to bring about even more sweeping worldwide ecological changes than we have seen in the past.

The idea of looking at cultural development in terms of a few relatively abrupt "revolutions" is obviously an oversimplification. Yet equally obviously, man's history does not show a smooth and gradual increase in knowledge and power. Shifts in techniques and ideas have resulted in periods of rapid change; and there have been long periods in the history of every culture in which neither ideas nor ways of life have changed much—periods of temporary near-equilibrium, or "stagnation" as some would say. The charting and description of these developments are matters for historians; and, although they may never reach agreement, their efforts are nonetheless thought-provoking in the attempts that each of us make to understand ourselves and our world.

Whatever the history, whatever the interplay of cause and effect, the result is the curious paradox of man as a part of nature, and man as apart from nature—a force of geological magnitude changing the face of the earth. Many of the activities that we think of as peculiarly human have parallels in other animals, but man has come to work on a different scale. Compare, for instance, Hoover Dam with the work of beavers, or Manhattan Island with a gopher town. The difference in scale is so great that we can only regard it as a difference in kind. In a somewhat comparable way, the human ecosystem has become a different sort of phenomenon from anything else that we know.

The world of life—the biosphere—can be looked at as a single, interconnected though endlessly diversified system. There is no denying the diversity. A desert and a forest are different kinds of places however difficult it may be to delineate their boundaries. Yet the many different kinds of biological communities that make up the biosphere display a very real underlying unity. Life everywhere is organized on the same basic principles.

Essentially all life as we know it depends on the transformation of radiant energy from the sun into the chemical energy of hydrocarbons through the process of photosynthesis based on chlorophyll. The chlorophyll-bearing organisms of the seas are mostly microscopic floating algae of the surface waters where sunlight penetrates. Such algae are probably also more important in fresh water than are the fixed plants of the shallows. But on land the "vascular" plants, the herbs, grasses, shrubs, and trees, are overwhelmingly important. These organisms that carry on the photosynthetic process are called *producers* by ecologists. The animals that live off them directly or indirectly are the *consumers*.

The British ecologist Charles Elton[4] has called the animals that live directly off the plants "key industry animals," because all the rest of the animal system depends on them. They can also be called "first-order consumers," which in turn are eaten by "second-order consumers," and so on. This leads to the idea of a food chain—grass, grasshoppers, frogs, snakes, hawks—which is a neat idea, but vastly oversimplified, because any attempt to diagram who eats whom in a biological community results in a complicated network of lines more appropriately called a "food web."

Most ecologists would add a third category of organisms, the *decomposers:* bacteria, fungi, and the like, that cause rot and decay, reducing the corpses of dead animals and plants to dust (or mud). The component chemicals are then available to be used again—carbon from the carbon dioxide of the air combined with oxygen from water through photosynthesis and the other necessary chemicals through absorption by plants or digestion by animals.

Most vegetable material never passes through the animal system at all, as is obvious enough in a glance at the uneaten leaves of any forest or prairie. This economy of abundance is necessary for the functioning of the biological community. When, because of some upset in the balance of the system, the forest is defoliated or the prairie overgrazed, the consequence is catastrophe for both the plants and the animals. In general, throughout nature, animal populations do not multiply up to the limit of their food supply—other controlling factors intervene. Plants, however, under favorable circumstances, may occupy all available space.

Most of the key industry animals (first-order consumers) are small, and often individuals of a given species are very abundant—insects on land, and minute crustaceans in water. Yet some of the largest of animals are herbivores—the elephants of today and the giant mammals, reptiles, and even birds of times past. Generally, however, the herbivores are more numerous than the predators that feed on them; and various studies have shown that, with each step away from the producer plants, only six to ten percent of the available energy is transmitted. It is not possible for animals, either in terms of energy or numbers, to operate beyond the fourth- or fifth-order consumer level.

What limits the number of individuals of a given species of animal? In 1798, Thomas Robert Malthus published his little book, *An Essay on the Principle of Population as It Affects the Future Improvement of Mankind.*[5] Concern with the human population problem can be neatly dated from the publication of this book. An enlarged and much revised edition, published in 1803, states his basic propositions thus:

> *1. Population is necessarily limited by the means of subsistence.*
>
> *2. Population invariably increases where the means of subsistence increase, unless prevented by some very powerful and obvious checks.*
>
> *3. These checks, and the checks which repress the superior power of population and keep its effects on a level with the means of subsistence, are all resolvable into moral restraint, vice and misery.*

This has been called the "dismal theorem" of Malthus, and there are those who would exorcise it with ridicule. But, dismal though it may seem, its basic validity remains unaffected either by man's ingenuity in finding means for the increase of his own subsistence or the likelihood that he will continue to achieve such increases up to some limit not yet reached. The theorem, of course, was intended to apply to people: "moral restraint" and "vice" can hardly refer to any animal except man. Under "misery," however, Malthus included such things as disease and starvation, which are not restricted to humans. In any event, it is instructive to consider his propositions in the context of the behavior of animal populations in general and not just that of *Homo sapiens.* Food is certainly the ultimate limit on the population of any animal, whether we think in terms of local or global populations. Yet, as noted above, populations in nature very rarely multiply up to the limit of the food supply. What, then, are the checks that normally limit animal populations? What prevents a population from multiplying until it reaches the point at which its food supply is exhausted? The search for the answers to such questions has led to a great deal of experiment, observation, and speculation. It is clear that there is no single kind of check, no simple answer—that we are dealing with a complex system of checks and balances that we cannot yet describe completely, let alone fully understand.[6]

Perhaps the most general check is the so-called balance-of-nature resulting from the fact that animals not only eat but are eaten. Herbivores tend to be controlled by carnivores before they reach the point of exhausting food supply as has been amply suggested by the damaging multiplication of such animals as deer or rabbits when man has removed their "enemies" such as pumas, wolves, or foxes. The carnivores may be limited in turn by other carnivores that live on them. The young of all animals are particularly susceptible to predation and to death through accident. And the most lordly carnivore may fall victim to parasitic disease. Disease particularly tends to be "density-dependent," becoming more prevalent as individuals of a particular host species become more common.

The whole system has a great deal of flexibility, of "play." A given predator in a particular region may concentrate largely on some one species of prey until this becomes rare and hard to find, then shift its attention to other species. More commonly, a predator may eat a variety of animals; but since the most abundant kinds would be most often caught, there is a sort of automatic limit on abundance for any given species.

The availability of food, however, is not the only factor limiting population size. A population may be limited by the availability of suitable breeding and rearing sites or by other special requirements of habitat. We are gradually learning about a variety of mechanisms that operate within a population to limit its growth. These include such things as cannibalism and failure to reproduce because of crowding, endocrine stress, and so on. Perhaps the most extensively studied of these limiting factors is territoriality. With many kinds of animals, especially fishes, lizards, birds, and mammals, an individual, a pair, or a social group may inhabit a particular area and defend it against intrusion by other individuals of the same species. Different kinds of territoriality have been described. All serve to space individuals; but where the defended area is large enough to include more food than can be utilized by the defenders, territorial behavior keeps the number of individuals below the theoretical limit imposed by the means of subsistence. The limit thus can be availability of appropriate space as well as food or other controls.

There is thus a constant turnover of individuals within a biological community, but with mortality balanced by reproduction so that the result, over a period of time, is a fairly steady state—a kind of dynamic equilibrium in the functioning of the community. This is comparable to the equilibrium that is maintained by the body of an individual: cells dying and being replaced, various organ systems working together harmoniously through an elaborate system of nerve and endocrine controls. This tendency to maintain a steady state despite environmental changes, stresses, and shocks (called *homeostasis*) is expressed in various natural population controls and the resulting balance of nature.[7]

Ecologists find it convenient to use, as their unit of study, the *ecosystem* rather than only the biological community, thus taking into account both the living organisms and their physical environment, which together form an interacting system. Every organism is affected by the conditions of the world in which it lives, but every organism also has some effect on these conditions, however trivial. The kind of forest growing in a particular region is in part a consequence of the soil, climate, and water supply of that region; but the kind of soil is also in part a consequence of the type of forest—coniferous, hardwood, or other—and the presence of the forest has a measurable influence on local climatic conditions and water supply.

The reciprocal interactions between organisms and environment are particularly striking in the case of the human species. Our activities are influenced in many ways by the nature of the physical setting in which we live—coastal, inland, desert, mountain, or forest. But we are also capable of altering the environment with unprecedented speed and effect. We would do well, then, to think in terms not of modern man in dominant relationship with other biological communities but of the human ecosystem—of the man-altered landscape and its biological components.

Man's actions can be looked at as efforts to simplify the biological relationships within the ecosystem to his own advantage. By clearing land and planting crops or orchards a complex of mixed species of wild plants may be replaced by a single kind of plant, a monoculture, which may extend over a wide area. In living off grain or fruit or tubers, man functions as a first-order consumer. In largely vegetarian societies he is, thus, a "key industry animal," which means that a large population can be supported—but he is also a dead end, not giving support in turn to the usual predators and scavengers. With modern medicine he has even largely defeated the parasites.

Man's food web, in such societies, is thus reduced to a simple producer-consumer interaction, often with the decomposer system greatly modified as plant growth is maintained by adding fertilizers to the soil. Man also tries as far as possible to reduce or eliminate competition, controlling the insect pests of the crops and attempting to eliminate vertebrate competitors, whether crows, rats, or raccoons.

As a meat-eater, man becomes a second-order consumer, growing grain for his chickens or hogs or pasture for his cattle and then eating the animals. Again there is an attempt to eliminate competition from hawks, weasels, big cats, or wolves; again there is a vastly simplified food relationship, unique to the human ecosystem.

These simplified food relationships are efficient as long as they can be sustained, and they now support the large and expanding human population. In fact, they could be regarded (in conjunction with medical reduction of the death rate) as a cause of this population growth, in terms of the second of the Malthusian propositions. There is no doubt about the effectiveness of the system to support an exploding human population *up to a point,* but it carries the danger of inherent imbalance and some large questions of ethics and aesthetics.

The danger in the simplified ecosystem is in its liability to catastrophe.[8] The most complex of natural ecosystems (in the sense of those including the largest numbers of different kinds of organisms) are the most stable in that they are the least liable to great fluctuations in number of individuals of a particular species from year to year. Greatest contrast observed is between complex biotas such as those of the tropical rain forest or coral reef and the relatively simple biotas of the arctic tundra or northern forests. Records of fluctuations in various tundra animals suggest either a cyclic Malthusian relationship with food supply or catastrophic physiological or psychological controls. Such cyclic fluctuations are unknown in the rain forest, where the complexity of relations makes for flexibility and results in a relatively steady state among limited populations of the many species there.

Charles Elton[9] has summarized six lines of evidence for the relative stability of complex communities. First there is the mathematical argument: "models" of simple predator-prey population relations show conspicuous fluctuations. As Elton says, "Put in ordinary language, this means that an animal community with only two such species in it would never have constant population levels, but would be subject to periodic 'outbreaks' of each species." The second line of evidence comes from attempts to test these simple relationships under laboratory conditions. "One thing stands out from the results: it is very difficult to keep small populations of this simple mixture in balance, for not only do they fluctuate but one or both of the species is liable to become extinct." Greater stability can be achieved by arranging for complex experimental conditions that provide cover for the prey to hide and dodge about in; but it is still difficult to maintain prey-predator populations for any length of time.

"The third piece of evidence," to quote Elton again, "is that natural habitats on small islands seem to be much more vulnerable to invading species than those of the continents." Animals and plants, accidentally or intentionally brought into simple island communities by man, often undergo catastrophic population explosions. Hawaii has seen many examples of this. The efforts at cure have involved attempts to restore a steady state by increasing the complexity of the system through the introduction of parasites and predators of the invading aliens.

"The fourth point is that invasions and outbreaks most often happen on cultivated or planted land—that is, in habitats and communities very much simplified by man." Our crops are notoriously subject to damage by pests, both native and foreign, which must be kept in control by chemical means —the chemicals sometimes producing unexpected side effects.

Elton's fifth line of evidence is based on the contrast in stability between tropical and arctic communities, as mentioned above. His final point involves recent research on orchard pest control. "Orchards are especially good for testing the effects of ecological variety, because they are half-way between a natural woodland and an arable field crop—less complex than the wood but more complex than the crop, and more permanent." Attempts to control orchard pests have disclosed many unexpected relationships among the populations involved. Elton quotes an observer as remarking: "We move from crisis to crisis, merely trading one problem for another."

Analogy can be made between the value of diversified ecosystems and the value of diversified national economies: where the economy depends largely on a single crop or a single industry, the danger of disruption is always greater than in situations where the economy has a diversified base. Large areas devoted to a single crop, covered by a single species of plant, or inhabited by concentrated numbers of a particular kind of animal, always present conditions conducive to the development of epidemics. History affords many examples. One of the most dramatic and best documented involves the Irish potato blight of 1845.[10] The Irish population provided a neat demonstration of the validity of the Malthusian propositions within something like a closed system by growing rapidly, following introduction of the potato, to more than eight million at the time of the census of 1841. When the crop failed in 1845 and again in 1846 because of the sudden appearance of the blight, more than a million people died directly of starvation, and as many as could emigrated. The shock in this case resulted in an eventual stabilization of the Irish population, which has remained at a level of about four million for the last fifty years—half of the preblight figure. In Malthusian terms, "moral restraint" (aided by migration) has checked the tendency to multiply up to the limit of the means of subsistence again.

Modern man, thus, depends on the simplified food relations of intensive agriculture, and modern technology has been efficient in developing methods of pest control. But the danger is always there—of a new pest, of immunity to chemical methods of control, and of environmental pollution from the pesticides. It would seem most prudent, therefore, to preserve as much of the natural diversity as possible, as is done in densely settled Europe with the hedgerows—at least until we understand the system well enough to foresee, guard against, and repair the consequences of possible oversimplification.

The aesthetic argument depends on the impression that a varied landscape is more pleasing, more satisfactory for living, then a monotonous one. This may be a matter for debate; and it carries over from landscapes to the general question of the value of diversity in styles of life, in ways of thinking and acting. From diversity comes the possibility of change, of adaptive response to new conditions, of development and evolution.

There would thus seem to be a need, in planning, to maintain as much diversity as possible within the human habitat. There is also a need to preserve substantial areas in which disturbance by man is kept at a minimum —the concept of nature reservations and national parks. Such parks serve at least four different and at times conflicting purposes: the recreation function, providing the possibility of escaping temporarily from the all-encompassing man-altered landscape; the research function, allowing the continued study of natural ecosystems; the museum function, preserving for future generations adequate samples of the diversified biosphere that has developed on our planet; and, perhaps most important of all, a reservoir function, whereby are sustained the organisms from which destructively altered environments may be restored.

The ethical question has been little explored by our philosophers. We have the power to alter the landscape and to exterminate species of animals and plants that we do not like or that we do not find convenient to our purposes. But do we have the right? Geological history is, of course, in part a record of the disappearance and replacement of organic types. But the rate of extinction increased enormously following the evolution of man as an ecologic dominant—and the rate is accelerating with the growth of human numbers, human power, and human intervention. Elton mentions a remark from Albert Schweitzer's book *Out of My Life and Thought*, pointing out that a fault of ethical philosophy has been that it has dealt only with the relations of man to man. And Aldo Leopold has decried the same oversight in terms of our need to develop an "ecological conscience."

Problems of immediate practicality and utility may override such aesthetic and ethical considerations locally or in emergencies; but the three in large measure coincide. There is need periodically to take stock of and to reconsider our behavior in terms of our own survival. As the demographers have shown, we are rapidly multiplying toward the spatial limits of our earth and toward a possibly tragic final test of the truth of the dismal propositions developed by Malthus. Our industrial society is not sufficiently aware of its needs to recycle the very materials on which its survival depends. We are changing the composition of air, water, and soil in ways that are to some extent irreversible and that have already in places reached disastrous proportions.

In very large measure, we have the knowledge we need to live harmoniously within the confines of our "Spaceship Earth"[11] if we do not insist on the constant growth of populations and new technology. The question that looms threateningly over the future of our species is, will we gain the wisdom to use this knowledge and to generate the new knowledge needed to evaluate and to cope with the environmental consequences of advancing technology?

... The needed wisdom must come from the people, who will demand action, and from their governments, which will enact the legislation and administer the programs needed to stabilize populations at levels that can be supported and to regulate the use of resources with the needs of the future in view.

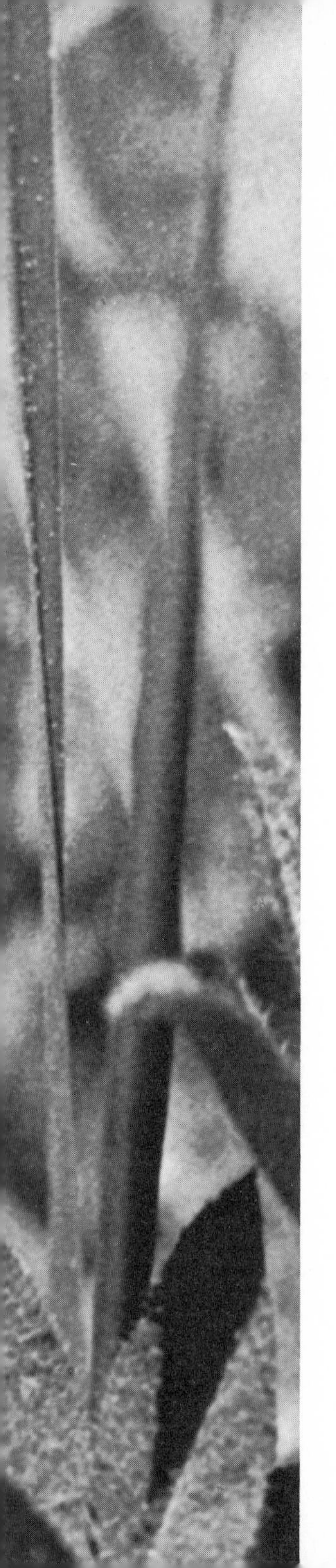

Fewer than thirty years have passed since the introduction of DDT as a pesticide, yet today measurable amounts of DDT are present in the fatty tissues of every man, woman, child, and animal on the earth, from the Eskimos of Northern Alaska to the penguins of the Antarctic. Both the Rienows and Marston Bates mentioned the movement of toxic substances like DDT through food chains. In the following article, George M. Woodwell shows how radioactive elements and pesticides which are released into the environment not only move through biological and meteorological cycles, but may also become concentrated in dangerous and unpredictable ways. His call for new, nonpoisonous methods of insect control echoes that of Rachel Carson in her essay, "The Other Road" (page 283) George M. Woodwell (1928-) received his doctorate in botany from Duke University and is a professional ecologist.

TOXIC SUBSTANCES and ECOLOGICAL CYCLES

by George M. Woodwell

The vastness of the earth has fostered a tradition of unconcern about the release of toxic wastes into the environment. Billowing clouds of smoke are diluted to apparent nothingness; discarded chemicals are flushed away in rivers; insecticides "disappear" after they have done their job; even the massive quantities of radioactive debris of nuclear explosions are diluted in the apparently infinite volume of the environment. Such pollutants are indeed diluted to traces—to levels infinitesimal by ordinary standards measured as parts per billion or less in air, soil and water. Some pollutants do disappear; they are immobilized or decay to harmless substances. Others last, sometimes in toxic form, for long periods. We have learned in recent years that dilution of persistent pollutants even to trace levels detectable only by refined techniques is no guarantee of safety. Nature has ways of concentrating substances that are frequently surprising and occasionally disastrous.

We have had dramatic examples of one of the hazards in the dense smogs that blanket our cities with increasing frequency. What is less widely realized is that there are global, long-term ecological processes that concen-

---

trate toxic substances, sometimes hundreds of thousands of times above levels in the environment. These processes include not only patterns of air and water circulation but also a complex series of biological mechanisms. Over the past decade detailed studies of the distribution of both radioactive debris and pesticides have revealed patterns that have surprised even biologists long familiar with the unpredictability of nature.

Major contributions to knowledge of these patterns have come from studies of radioactive fallout. The incident that triggered worldwide interest in large-scale radioactive pollution was the hydrogen-bomb test at Bikini in 1954 known as "Project Bravo." This was the test that inadvertently dropped radioactive fallout on several Pacific islands and on the Japanese fishing vessel *Lucky Dragon*. Several thousand square miles of the Pacific were contaminated with fallout radiation that would have been lethal to man. Japanese and U.S. oceanographic vessels surveying the region found that the radioactive debris had been spread by wind and water, and, more disturbing, it was being passed rapidly along food chains from small plants to small marine organisms that ate them to larger animals (including the tuna, a staple of the Japanese diet).

The U.S. Atomic Energy Commission and agencies of other nations, particularly Britain and the U.S.S.R., mounted a large international research program, costing many millions of dollars, to learn the details of the movement of such debris over the earth and to explore its hazards. Although these studies have been focused primarily on radioactive materials, they have produced a great deal of basic information about pollutants in general. The radioactive substances serve as tracers to show the transport and concentration of materials by wind and water and the biological mechanisms that are characteristic of natural communities.

One series of investigations traced the worldwide movement of particles in the air. The tracer in this case was strontium 90, a fission product released into the earth's atmosphere in large quantities by nuclear-bomb tests. Two reports in 1962—one by S. Laurence Kulp and Arthur R. Schulert of Columbia University and the other by a United Nations committee—furnished a detailed picture of the travels of strontium 90. The isotope was concentrated on the ground between the latitudes of 30 and 60 degrees in both hemispheres, but concentrations were five to 10 times greater in the Northern Hemisphere, where most of the bomb tests were conducted.

It is apparently in the middle latitudes that exchanges occur between the air of upper elevations (the stratosphere) and that of lower elevations (the troposphere). The larger tests have injected debris into the stratosphere; there it remains for relatively long periods, being carried back into the troposphere and to the ground in the middle latitudes in late winter or spring. The mean "half-time" of the particles' residence in the stratosphere (that is, the time for half of a given injection to fall out) is from three months to five years, depending on many factors, including the height of the injection, the size of the particles, the latitude of injection and the time of year. Debris injected into the troposphere has a mean half-time of residence ranging from a few days to about a month. Once airborne, the particles may travel rapidly and far. The time for one circuit around the earth in the middle latitudes varies from 25 days to less than 15. (Following two recent bomb tests in China fallout was detected at the Brookhaven National Laboratory on Long Island respectively nine and 14 days after the tests.)

Numerous studies have shown further that precipitation (rain and snowfall) plays an important role in determining where fallout will be deposited. Lyle T. Alexander of the Soil Conservation Service and Edward P. Hardy, Jr., of the AEC found in an extensive study in Clallam County, Washington, that the amount of fallout was directly proportional to the total annual rainfall.

It is reasonable to assume that the findings about the movement and fallout of radioactive debris also apply to other particles of similar size in the air. This conclusion is supported by a recent report by Donald F. Gatz and A. Nelson Dingle of the University of Michigan, who showed that the concentration of pollen in precipitation follows the same pattern as that of radioactive fallout. This observation is particularly meaningful because pollen is not injected into the troposphere by a nuclear explosion; it is picked up in air currents from plants close to the ground. There is little question that dust and other particles, including small crystals of pesticides, also follow these patterns.

From these and other studies it is clear that various substances released into the air are carried widely around the world and may be deposited in concentrated form far from the original source. Similarly, most bodies of water—especially the oceans—have surface currents that may move materials five to 10 miles a day. Much higher rates, of course, are found in such major oceanic currents as the Gulf Stream. These currents are one more physical mechanism that can distribute pollutants widely over the earth.

The research programs of the AEC and other organizations have explored not only the pathways of air and water transport but also the pathways along which pollutants are distributed in plant and animal communities. In this connection we must examine what we mean by a "community."

Biologists define communities broadly to include all species, not just man. A natural community is an aggregation of a great many different kinds of organisms, all mutually interdependent. The basic conditions for the integration of a community are determined by physical characteristics of the environment such as climate and soil. Thus a sand dune supports one kind of community, a freshwater lake another, a high mountain still another. Within each type of environment there develops a complex of organisms that in the course of evolution becomes a balanced, self-sustaining biological system.

Such a system has a structure of interrelations that endows the entire community with a predictable developmental pattern, called "succession," that leads toward stability and enables the community to make the best use of its physical environment. This entails the development of cycles through which the community as a whole shares certain resources, such as mineral nutrients and energy. For example, there are a number of different inputs of nutrient elements into such a system. The principal input is from the decay of primary minerals in the soil. There are also certain losses, mainly through the leaching of substances into the underlying water table. Ecologists view the cycles in the system as mechanisms that have evolved to conserve the elements essential for the survival of the organisms making up the community.

One of the most important of these cycles is the movement of nutrients and energy from one organism to another along the pathways that are sometimes called food chains. Such chains start with plants, which use the sun's energy to synthesize organic matter; animals eat the plants; other animals eat these herbivores, and carnivores in turn may constitute additional levels feeding on the herbivores and on one another. If the lower orders in the chain are to survive and endure, there must be a feedback of nutrients. This is provided by decay organisms (mainly microorganisms) that break down organic debris into the substances used by plants. It is also obvious that the community will not survive if essential links in the chain are eliminated; therefore the preying of one level on another must be limited.

Ecologists estimate that such a food chain allows the transmission of roughly 10 percent of the energy entering one level to the next level above it, that is, each level can pass on 10 percent of the energy it receives from below without suffering a loss of population that would imperil its survival. The simplest version of a system of this kind takes the form of a pyramid, each successively higher population receiving about a tenth of the energy received at the level below it.

Actually nature seldom builds communities with so simple a structure. Almost invariably the energy is not passed along in a neatly ordered chain but is spread about to a great variety of organisms through a sprawling, complex web of pathways. The more mature the community, the more diverse its makeup and the more complicated its web. In a natural ecosystem the network may consist of thousands of pathways.

This complexity is one of the principal factors we must consider in investigating how toxic substances may be distributed and concentrated in living communities. Other important basic factors lie in the nature of the metabolic process. For example, of the energy a population of organisms receives as food, usually less than 50 percent goes into the construction of new tissue, the rest being spent for respiration. This circumstance acts as a concentrating mechanism: a substance not involved in respiration and not excreted efficiently may be concentrated in the tissues twofold or more when passed from one population to another.

Let us consider three types of pathway for toxic substances that involve man as the ultimate consumer. The three examples, based on studies of radioactive substances, illustrate the complexity and variety of pollution problems.

The first and simplest case is that of strontium 90. Similar to calcium in chemical behavior, this element is concentrated in bone. It is a long-lived radioactive isotope and is a hazard because its energetic beta radiation can damage the mechanisms involved in the manufacture of blood cells in the bone marrow. In the long run the irradiation may produce certain types of cancer. The route of strontium 90 from air to man is rather direct: we ingest it in leafy vegetables, which absorbed it from the soil or received it as fallout from the air, or in milk and other dairy products from cows that have fed on contaminated vegetation. Fortunately strontium is not usually concentrated in man's food by an extensive food chain. Since it lodges chiefly in bone, it is not concentrated in passing from animal to animal in the same ways other radioactive substances may be (unless the predator eats bones!).

Quite different is the case of the radioactive isotope cesium 137. This isotope, also a fission product, has a long-lived radioactivity (its half-life is about 30 years) and emits penetrating gamma rays. Because it behaves chemically like potassium, an essential constituent of all çells, it becomes widely distributed once it enters the body. Consequently it is passed along to meat-eating animals, and under certain circumstances it can accumulate in a chain of carnivores.

A study in Alaska by Wayne C. Hanson, H. E. Palmer and B. I. Griffin of the AEC's Pacific-Northwest Laboratory showed that the concentration factor for cesium 137 may be two or three for one step in a food chain. The first link of the chain in this case was lichens growing in the Alaskan forest and tundra. The lichens collected cesium 137 from fallout in rain. Certain caribou in Alaska live mainly on lichens during the winter, and caribou meat in turn is the principal diet of Eskimos in the same areas. The investigators found that caribou had accumulated about 15 micromicrocuries of cesium radioactivity per gram of tissue in their bodies. The Eskimos who fed on these caribou had a concentration twice as high (about 30 micromicrocuries per gram of tissue) after eating many pounds of caribou meat in the course of a season. Wolves and foxes that ate caribou sometimes contained three times the concentration in the flesh of the caribou. It is easy to see that in a longer chain, involving not just two animals but several, the concentration of a substance that was not excreted or metabolized could be increased to high levels.

A third case is that of iodine 131, another gamma ray emitter. Again the chain to man is short and simple: The contaminant (from fallout) comes to man mainly through cows' milk, and thus the chain involves only grass, cattle, milk and man. The danger of iodine 131 lies in the fact that iodine is concentrated in the thyroid gland. Although iodine 131 is short-lived (its half-life is only about eight days), its quick and localized concentration in the thyroid can cause damage. For instance, a research team from the Brookhaven National Laboratory headed by Robert Conard has discovered that children on Rongelap Atoll who were exposed to fallout from the 1954 bomb test later developed thyroid nodules.

The investigations of the iodine 131 hazard yielded two lessons that have an important bearing on the problem of pesticides and other toxic substances released in the environment. In the first place we have had a demonstration that the hazard of the toxic substance itself often tends to be underestimated. This was shown to be true of the exposure of the thyroid to radiation. Thyroid tumors were found in children who had been treated years before for enlarged thymus glands with doses of X rays that had been considered safe. As a result of this discovery and studies of the effects of iodine 131, the Federal Radiation Council in 1961 issued a new guide reducing the permissible limit of exposure to ionizing radiation to less than a tenth of what had previously been accepted. Not the least significant aspect of this lesson is the fact that the toxic effects of such a hazard may not appear until long after the exposure; on Rongelap Atoll 10 years passed before the thyroid abnormalities showed up in the children who had been exposed.

The second lesson is that, even when the pathways are well understood, it is almost impossible to predict just where toxic substances released into the environment will reach dangerous levels. Even in the case of the simple pathway followed by iodine 131 the eventual destination of the substance and its effects on people are complicated by a great many variables: the area of the cow's pasture (the smaller the area, the less fallout the cow will pick up); the amount and timing of rains on the pasture (which on the one hand may bring down fallout but on the other may wash it off the forage); the extent to which the cow is given stored, uncontaminated feed; the amount of iodine the cow secretes in its milk; the amount of milk in the diet of the individual consumer, and so on.

If it is difficult to estimate the nature and extent of the hazards from radioactive fallout, which have been investigated in great detail for more than a decade by an international research program, it must be said that we are in a poor position indeed to estimate the hazards from pesticides. So far the amount of research effort given to the ecological effects of these poisons has been comparatively small, although it is increasing rapidly. Much has been learned, however, about the movement and distribution of pesticides in the environment, thanks in part to the clues supplied by the studies of radioactive fallout.

Our chief tool in the pesticide inquiry is DDT. There are many reasons for focusing on DDT: it is long-lasting, it is now comparatively easy to detect, it is by far the most widely used pesticide and it is toxic to a broad spectrum of animals, including man. Introduced only a quarter-century ago and spectacularly successful during World War II in controlling body lice and therefore typhus, DDT quickly became a universal weapon in agriculture and in public health campaigns against disease-carriers. Not surprisingly, by this time DDT has thoroughly permeated our environment. It is found in the air of cities, in wildlife all over North America and in remote corners of the earth, even in Adélie penguins and skua gulls (both carnivores) in the

Antarctic. It is also found the world over in the fatty tissue of man. It is fair to say that there are probably few populations in the world that are not contaminated to some extent with DDT.

We now have a considerable amount of evidence that DDT is spread over the earth by wind and water in much the same patterns as radioactive fallout. This seems to be true in spite of the fact that DDT is not injected high into the atmosphere by an explosion. When DDT is sprayed in the air, some fraction of it is picked up by air currents as pollen is, circulated through the lower troposphere and deposited on the ground by rainfall. I found in tests in Maine and New Brunswick, where DDT has been sprayed from airplanes to control the spruce budworm in forests, that even in the open, away from trees, about 50 percent of the DDT does not fall to the ground. Instead it is probably dispersed as small crystals in the air. This is true even on days when the air is still and when the low-flying planes release the spray only 50 to 100 feet above treetop level. Other mechanisms besides air movement can carry DDT for great distances around the world. Migrating fish and birds can transport it thousands of miles. So also do oceanic currents. DDT has only a low solubility in water (the upper limit is about one part per billion), but as algae and other organisms in the water absorb the substance in fats, where it is highly soluble, they make room for more DDT to be dissolved into the water. Accordingly water that never contains more than a trace of DDT can continuously transfer it from deposits on the bottom to organisms.

DDT is an extremely stable compound that breaks down very slowly in the environment. Hence with repeated spraying the residues in the soil or water basins accumulate. Working with Frederic T. Martin of the University of Maine, I found that in a New Brunswick forest where spraying had been discontinued in 1958 the DDT content of the soil increased from half a pound per acre to 1.8 pounds per acre in the three years between 1958 and 1961. Apparently the DDT residues were carried to the ground very slowly on foliage and decayed very little. The conclusion is that DDT has a long half-life in the trees and soil of a forest, certainly in the range of tens of years.

Doubtless there are many places in the world where reservoirs of DDT are accumulating. With my colleagues Charles F. Wurster, Jr., and Peter A. Isaacson of the State University of New York at Stony Brook, I recently sampled a marsh along the south shore of Long Island that had been sprayed with DDT for 20 years to control mosquitoes. We found that the

DDT residues in the upper layer of mud in this marsh ranged up to 32 pounds per acre!

We learned further that plant and animal life in the area constituted a chain that concentrated the DDT in spectacular fashion. At the lowest level the plankton in the water contained .04 part per million of DDT; minnows contained one part per million, and a carnivorous scavenging bird (a ring-billed gull) contained about 75 parts per million in its tissues (on a whole-body, wet-weight basis). Some of the carnivorous animals in this community had concentrated DDT by a factor of more than 1,000 over the organisms at the base of the ladder.

A further tenfold increase in the concentrations along this food web would in all likelihood result in the death of many of the organisms in it. It would then be impossible to discover why they had disappeared. The damage from DDT concentration is particularly serious in the higher carnivores. The mere fact that conspicuous mortality is not observed is no assurance of safety. Comparatively low concentrations may inhibit reproduction and thus cause the species to fade away.

That DDT is a serious ecological hazard was recognized from the beginning of its use. In 1946 Clarence Cottam and Elmer Higgins of the U.S. Fish and Wildlife Service warned in the *Journal of Economic Entomology* that the pesticide was a potential menace to mammals, birds, fishes and other wildlife and that special care should be taken to avoid its application to streams, lakes and coastal bays because of the sensitivity of fishes and crabs. Because of the wide distribution of DDT the effects of the substance on a species of animal can be more damaging than hunting or the elimination of a habitat (through an operation such as dredging marshes). DDT affects the entire species rather than a single population and may well wipe out the species by eliminating reproduction.

Within the past five years, with the development of improved techniques for detecting the presence of pesticide residues in animals and the environment, ecologists have been able to measure the extent of the hazards presented by DDT and other persistent general poisons. The picture that is emerging is not a comforting one. Pesticide residues have now accumulated to levels that are catastrophic for certain animal populations, particularly carnivorous birds. Furthermore, it has been clear for many years that because of their shotgun effect these weapons not only attack the pests but also destroy predators and competitors that normally tend to limit proliferation of the pests. Under exposure to pesticides the pests tend to develop new strains that are resistant to the chemicals. The result is an escalating chemical warfare that is self-defeating and has secondary effects whose costs are only beginning to be measured. One of the costs is wildlife, notably carnivorous and scavenging birds such as hawks and eagles. There are others: destruction of food webs aggravates pollution problems, particularly in bodies of water that receive mineral nutrients in sewage or in water draining from heavily fertilized agricultural lands. The plant populations, no longer consumed by animals, fall to the bottom to decay anaerobically, producing hydrogen sulfide and other noxious gases, further degrading the environment.

The accumulation of persistent toxic substances in the ecological cycles of the earth is a problem to which mankind will have to pay increasing attention. It affects many elements of society, not only in the necessity for concern about the disposal of wastes but also in the need for a revolution in pest control. We must learn to use pesticides that have a short half-life in the environment—better yet, to use pest-control techniques that do not require applications of general poisons. What has been learned about the dangers in polluting ecological cycles is ample proof that there is no longer safety in the vastness of the earth.

Population Explosion

Population Explosion

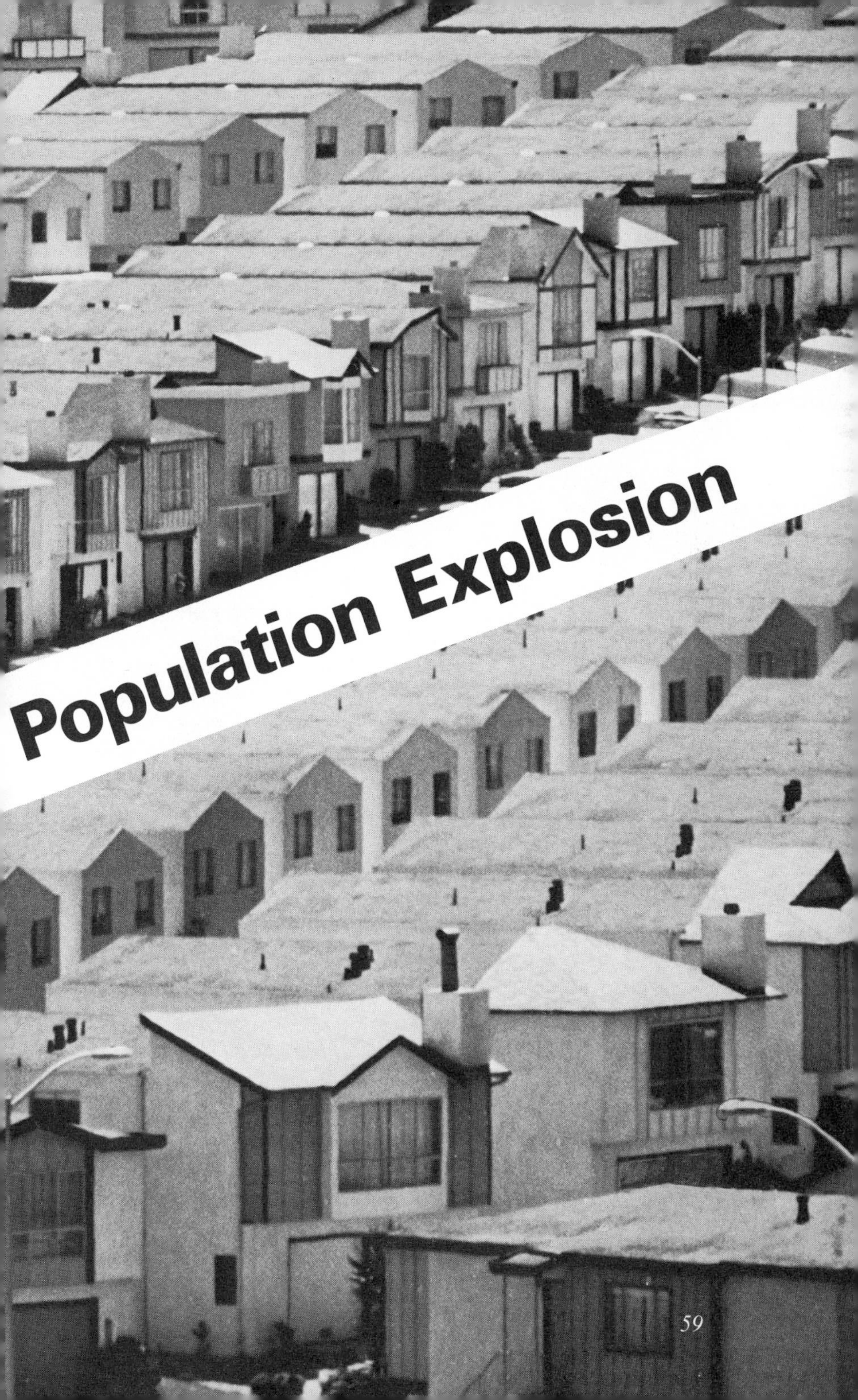

There can be little doubt that overpopulation is one of the major crises facing mankind today. No real progress can be made toward solving the earth's environmental, social, and economic problems until ways can be found to reduce the population growth rate. Before going on to deal with other ecological problems, it is important to understand more clearly the population crisis.

Sir Julian Huxley (1887-   ) examines the population "explosion" from both a historical and a current viewpoint. In Biblical times the world population doubled once every thousand years. According to latest figures, the world population today is doubling once every thirty-five years. As Huxley points out, the whole of mankind's future is at stake if this growth rate is not checked. The latter part of Huxley's essay is a discussion of possible solutions and a consideration of the ultimate philosophical question: What is the purpose of human life? Is the meaning of human existence to be measured by the numbers of people who can exist on the earth or by the quality of the lives which they lead?

It is not surprising that such philosophical questions preoccupy Huxley. He comes from an intellectually distinguished family. His grandfather was the famous biologist, Thomas Henry Huxley, and he is the elder brother of the late novelist, Aldous Huxley. He is an honor graduate in zoology from Oxford and is the author of many works in biology. He was knighted by Queen Elizabeth II in 1958.

THE WORLD POPULATION PROBLEM

by Sir Julian Huxley

The world population problem is to my mind the most important and the most serious of all the problems now besetting the human species. The problem of avoiding nuclear war is more immediate, but that of overpopulation is, in the long run, more serious and more difficult to deal with because it is rooted in our own nature.

It is now a world problem in general public estimation, but it has only recently achieved that position. It is striking to see the way in which it has suddenly emerged into public consciousness during the last few years. Let me remind you of its history. It was first posed by the Reverend Mr. Malthus well over a century ago. As a biologist, I like to recall that it was Malthus' work that led both Darwin and Wallace, independently, to the idea of natural selection. Darwin, for instance, records in his *Autobiography* how, during his voyage on H.M.S. *Beagle,* he had realized that evolution must be a fact, and how, after his return to England, he had realized from his studies of what man had done with breeds of domestic animals and strains of crop plants how deliberate artificial selection could modify and transform species of animals and plants. But he still could not see how anything like this

---

selective process could be operative in nature. But one day in 1838, as he records in a charming phrase, "I happened to read for amusement 'Malthus on Population.'" And as a result, the idea of natural selection flashed into his mind. Malthus' point was that population tends to grow in geometrical ratio, at a compound interest rate, whereas the means of supporting that population tend to grow only at a much lower rate. Though his gloomy prophecies were based on correct principles, they were temporarily falsified by two facts: just at this time intensive industrialization was leading to higher productivity, and the New World was opening up to immigration from the Old World. Accordingly, the disastrous pressure of population on resources which he foretold did not immediately occur. But it was only put off, and it is now occurring.

Next let me recall the extraordinary change in public attitude toward the idea of deliberately controlling population increase. Malthus was a clergyman and merely advocated what he called moral restraint. It speedily became apparent that, human nature being what it is, moral restraint certainly was not going to solve the population problem, and a few daring spirits began to advocate mechanical or chemical methods of preventing conception. But for many years the more orthodox members of the community regarded any idea of artificial birth control as immoral and indeed wicked.

I recently came across the astonishing fact that in the year 1873 Mr. Gladstone, the great Liberal statesman of Britain, withdrew his name from support of a memorial to John Stuart Mill because he had just discovered that John Stuart Mill had once advocated birth control. A little later came the famous Bradlaugh-Besant trial, when Annie Besant and Charles Bradlaugh were tried and sentenced for advocating artificial birth control. By the way, it is interesting to note that the publicity for the idea of birth control achieved by the trial had the result of changing the sign of the birth rate in Britain from positive to negative. Before this time the birth rate had been steadily going up; one year after the trial it reached its peak and began to go down.

Then in the early years of this century your great pioneer, Margaret Sanger, who I am glad to say is still alive, was jailed in New York for advocating birth control, and was actually arrested and jailed for a night in Portland, Oregon, for the same reason as late as 1921. I myself have also suffered from this same attitude. In 1929, or thereabouts, I advocated birth control in a radio talk about population on the B.B.C. in England. Sir John Reith (now Lord Reith), who was then the head of the B.B.C., actually sent for me as if I were a schoolboy and reprimanded me for having said something so terrible over the British ether.

That was only thirty-three years ago. Since then there has been an extraordinary breakthrough. A couple of years ago I was much honored by being given the Lasker award for my work in regard to population and had to make a speech on the subject on the occasion of receiving the award. I was warned that, owing to the general climate of opinion in the United States, I ought to be very tactful, and in preparing my speech I took a great deal of trouble to ensure that I was tactful. However, I need not have bothered. In the few weeks before the speech was actually delivered, both *Time* and *Life* magazines had comprehensive articles on the subject; two official committees, one presidential, the other senatorial, recommended that the United States ought to undertake much more research on reproduction and its control; and one of them added that the results of research ought to be made available to other nations on request. Then, of course, population control became quite an issue in the last presidential election, and today you can hardly open a newspaper without seeing some reference to the difficulties caused by excessive population growth in some part of the world or other. There really has been a breakthrough. The general public is now aware that population is a grave problem.

Originally, birth control was advocated almost entirely by women in the interest of women. It was advocated to save women from undesired pregnancies and unwanted children and from the illness and frustration that arise from excessive childbearing. It was in a sense a campaign against male selfishness. As such it was largely led by women—Mrs. Besant, Margaret Sanger, Marie Stopes, and others. Then it began to be seen that it was not merely a problem of the individual woman, the individual wife and mother; it was a problem of society as a whole, a problem affecting entire nations. It began to be very pressing, for instance, in countries like India. Eventually the movement changed its name. The British National Birth Control Association, founded in 1929, later became the Family Planning Association, while in 1939 the American Birth Control League was transformed into the Planned Parenthood Federation of America.

It was soon realized that there were great differences in the rate of increase between different nations. Western European countries were increasing only slowly (France was even perhaps on the verge of decreasing), whereas India and Japan and Java and some Latin-American countries were increasing very fast. The problem was obviously an international one. In

1927 the first World Population Conference was held in Geneva, organized by Margaret Sanger and her husband: I have vivid memories of it. In the same year Professor Raymond Pearl founded the International Union for the Scientific Investigation of Population Problems. Since then population has been increasingly recognized as a world problem, though of course with social and national, humanitarian, and individual aspects as well.

So much for history. What about the situation today? The situation is that we find ourselves caught up in what must be called a population explosion. This is something quite new. In man's first half million years the growth of the world's population must have been exceedingly slow. It seems certain that the whole population of mankind cannot have reached much more than ten million before the neolithic revolution resulting from the discovery of agriculture around 6000 B.C. Even after that the compound interest rate of human increase, what with infant mortality, famine, disease, and war, continued very low, probably not exceeding 0.1 per cent per annum. The growth of organized civilizations sent up the rate somewhat, but it was not until the middle of the seventeenth century A.D. (the first period for which we have any reasonable statistics) that the total population of the world reached about half a billion, and the rate of increase rose to nearly 0.5 per cent per annum.

Then came the great explorations, which opened up new lands for colonization, and the new industrial technology, which led to higher productivity. The total population of the world leapt up, and the rate of increase itself increased, until by the early 1900's it had reached nearly 1 per cent per annum, and there were approximately one and a half billion people in the world—a threefold increase in 250 years. After 1900 the process was again accelerated, so that by 1950 the rate of compound-interest increase had gone up to 1.5 per cent per annum. Today it is nearly 1.75 per cent and shows no sign of slowing down, while the total population of the world is over two and three-quarter billion. In the last eighty years the rate of doubling has itself doubled.

Let me spell out quantitatively what this means. Today, the annual net increase in the total of people on the world's surface is over fifty million. That means that the world's population grows by about 150,000 people every 24 hours, the equivalent of a medium-sized town every day, 365 of them every year. To bring the facts home to American audiences may I point out that this is the equivalent of ten baseball teams, complete with coach, every minute of every hour of every day. And yet there are people who seriously talk of exporting our surplus population to Mars or some other planet!

The single country, China, contains well over six hundred million people. A few months ago its net annual increase was stated to be over thirteen million—much larger than the *total* population of Australia and New Zealand combined—but the latest information indicates that it is really over fourteen million. This means that by the mid-1970's the *increase* in China's population will be greater than the present *total* population of the United States.

The United Nations Demographic Unit has gone into the world situation very carefully. It seems quite certain that by the year 1999—that is to say, while more than half the people now in existence on this planet will still be alive—there will certainly be well over five and a half billion people in the world, almost certainly over six billion, and quite possibly nearly seven billion. That is to say, the world population will about double itself within the next forty years.

Clearly this business of doubling cannot go on indefinitely, or indeed for more than a few decades, without leading to disaster. This is especially clear when we consider the *differential* rate of increase in different countries. The very high rates of increase are found mostly in the countries of Asia and in the tropical regions of Latin America and parts of Africa. The population explosion in these areas is undoubtedly due primarily to the great advances in medical science and its application in better health services. This has led to what has been pithily called "death control." Death control has been especially noticeable in regard to infant mortality. When I first went to Africa in 1929 I found to my surprise and horror that in many parts of East and Central Africa the infant mortality rates—the number of children dying the first year of life—ranged from 250 up to 500 per thousand: from one in four to one in two babies died before they reached the age of one. Today the rate has been drastically brought down: there are very few countries where infant mortality is one in four, and in many parts even of the tropical world it is down to one in eight or one in ten.

But death control operates at all ages, and people live longer. This you can measure by the increased expectation of life at birth. One of the most important factors in human history has been this increase in the expectation of life, even in the high civilizations. The inscriptions on Roman tombstones tell us that the average expectation of life in classical Rome at the height of Roman power was only about thirty years. Today in all developed countries it is more than twice this, and it has been going up in the underdeveloped countries, too. Professor Linton, the English geographer, has recently pointed out that in 1920 an average expectation of life of forty years or more occurred only in two temperate areas—Canada, the United States, and most of Europe in the northern hemisphere; the Argentine, part of South Africa, Australia, and New Zealand in the southern hemisphere. Only in one country in the world—New Zealand—was there an expectation of life of over sixty. Only thirty years later, in 1950, an expectation of life of forty or under was found only in North and Central Africa, in parts of the Middle East, in India, in most of China, and in parts of Latin America. In New Zealand, Australia, South Africa, all of northwest Europe, Canada, and the United States it was over sixty; in many parts it was over sixty-five; and in Scandinavia it had reached seventy—an incredible change in a mere thirty years.

The other great fact that has emerged into world consciousness is the enormous disparity in standards of life between the parts of the world that are overpopulated and underdeveloped and those that are less heavily populated and technologically better developed—the great gap between the have-nots and the haves, which is also the gap between the rapidly increasing and the slowly increasing countries. To take some examples, Asia includes one fifth of the world's land and nearly three fifths of the world's population. The rate of increase in western Europe is less than 1.5 per cent per annum, whereas the world increase is 1.75 per cent per annum and in many tropical areas it is well over 2 per cent. Some of the West Indian Islands and parts of the Indonesian Archipelago are even increasing at rates of 3 per cent and over, which means doubling their population in roughly twenty years: such increases are fantastic, and impossible to take care of. Furthermore, these rapid rates of increase occur where resources are least.

The contrast in resources and standards of living between the haves and the have-nots is very striking. Thus the real income per head per annum in the United States is over twenty times what it is in India. In regard to available energy and to steel consumption, the ratio is more like 40 to 1, and for things like newsprint it is probably in the neighborhood of 50 or even 100 to 1.

To less privileged nations, the consumption of newsprint in the United States seems ridiculously high, or at least selfishly excessive. To take a single example, every week the Sunday edition of the *New York Times* has more words in it than the entire Bible. And every year it takes a forest the size of Staffordshire to supply the pulp for it.

When we come to the most basic resource of all, namely food, the appalling fact emerges that over half the world's two and three-quarter billion people are not properly fed. They are hungry or malnourished in one way or another, incapable of achieving full growth or physical energy. They are also underhealthy, suffering from malaria or infectious disease. In addition, they are grossly undereducated. When I first came to Unesco in 1946, more than half the world's people were illiterate. Though the proportion has now gone down slightly, more than half the world's children still do not get even good primary education.

Not only is the gap between the haves and the have-nots large, but it is increasing: it is getting larger instead of smaller. On the other hand, the knowledge of the facts is spreading as the world shrinks and communications are improved. As a result, we are witnessing what has been called the revolution of expectation. The peoples of the underdeveloped countries have become aware of the huge differences in standards of living between different regions. They have also become aware that science can do wonders in the production of food and goods and in promoting better health. And they, quite rightly, expect that something will be done to raise their standard of living, in regard to food, health, education, and everything else. If this expectation is not fulfilled, at least to a reasonable degree, there will assuredly be an immense amount of discount, frustration, and unrest, even conflict and possibly war.

Now let me take some of the various ways in which the population explosion is now affecting human life, and affecting it in almost every case for the worse. First, then, the basic problem of food. Men must eat to live. In this field, many people are still thinking in terms of a race—a race between food and people, between production and reproduction. This, I am sure, is a wrong approach. The essence of a race is that it can be won; but this is something that neither side can ever win. The present rate of human increase cannot continue for more than a century or so at the outside without leading to a completely absurd but completely disastrous result. And this is for a period that is insignificant in relation to the evolutionary time before us—time to be measured not in hundreds, not in thousands, but in millions and even hundreds of millions of years. No, we must give up thinking in terms of a race; we must think in terms of a balance. We must aim at achieving some sort of balanced relation between the rate of reproduction and the rate of production of food and other resources.

It is often claimed that we could feed an enormously larger number of people if we used better methods of agriculture and if we cultivated the areas of the world's surface that are now unused. It is possible that we might be able to feed four or five times the present world population by a combination of improved methods and the exploitation of new areas, but it is not so simple as many people seem to think. The vast areas of tropical rain forest in Amazonia and Central Africa are extremely marginal for agriculture: the soil is very poor and, if the trees are cut down, and cultivation is attempted, it deteriorates rapidly. Much of the huge central area of Australia is desert or semidesert. There is negligible rainfall and no surface water; if you make bore holes, the water that comes up is often saline. And much of the region is covered by a veritable carapace of hard laterite instead of good soil.

In the vast savannah areas of Africa, the climate is difficult and the soil poor and likely to deteriorate under cultivation or with the slightest degree of overgrazing. Let me remind you that during the last few thousand years a large part of the world, even of the so-called civilized world, has been what I may call defertilized by human stupidity and greed. Most of the coastal belt of the eastern Mediterranean—northern Africa, Syria, Greece, what is now Yugoslavia—was once extremely fertile and acted as the granary of Rome. But men have cut down the forests, the soil has been eroded, the wildlife has been killed off, and now much of the area is semidesert and the rest of it highly unproductive. In some places attempts are being made to reforest and to restore fertility, but this is a long and expensive job.

You in America have had similar troubles. In the thirties there was the growth of the famous dust bowl: luckily American technical knowledge, skill, and good will were sufficient to overcome it. You had terrible erosion in parts of the South. When I was taken to see the Tennessee Valley area in 1933 or '34, just before the completion of the Norris Dam, I saw bare rock showing where some of the inhabitants could remember five feet of good soil. Earlier I had imagined that all American rivers were yellow and turbid, but I was told there were people still living who could remember the Tennessee River as a transparent, clear blue stream. What did this mean? It meant that year after year for many decades millions of tons of good soil had been washed down to sea, and the fertility of the land had been gradually exhausted. In this case, too, you have remedied this disastrous state of affairs, by means of that wonderful project, the T.V.A. But once more it shows how careful we must be if we are to conserve our precious resources instead of wasting them.

India, also, has suffered deforestation and a grave loss of soil fertility. The shortage of firewood has even made it necessary to use cow dung for fuel instead of returning it to the soil. It has been stated that an increased use of appropriate fertilizers could solve India's problem. Up to a point, this is true. If India could purchase an adequate amount of fertilizers and get them applied properly, the country could assuredly double its production of food. But this would not be easy. You have to pay for the fertilizer; you have to persuade the peasants to change their old habits, including their ingrained resistance to change. This will take time and a great deal of leadership and devoted effort. Finally, by the time you have doubled the production of food, the population too will have approximately doubled itself, so that at the end you will be hardly any further than you were at the beginning.

In British possessions, since the end of the war, there have been a number of attempts to raise food production intensively, but several of them have come to a disastrous end. The most famous, or perhaps I should say the most infamous, was the groundnut scheme in Tanganyika initiated by the Labor government in England just after the end of World War II. It seemed a wonderful idea. Groundnuts (or peanuts, as you call them) are a very valuable crop in Africa: for instance, they are the main basis of the prosperity of northern Nigeria. The climate of parts of Tanganyika seemed suitable, and vast areas were virtually unused. Why not exploit them for the mass production of groundnuts? Unfortunately, no proper ecological survey was made of the area, and no preliminary pilot projects were undertaken. The result, in a nutshell, was that the whole project had to be abandoned, and about thirty million pounds sterling went irrecoverably down the drain.

A similar fiasco, luckily not on such a massive scale but equally bad from the qualitative angle, occurred in the tiny West African colony of Gambia. The British government wanted to improve the economy of the colony by introducing poultry farming on a large scale. Unfortunately, here too they had not surveyed the situation adequately, notably as regards the climate and the proneness of the fowls to certain diseases, nor in their haste had they undertaken any pilot project. And again the result was that the whole scheme was a fiasco and had to be abandoned. My cousin by marriage, Elspeth Huxley, in her book about West Africa, mentions one curious aftermath of this disastrous project. When it was abandoned, the authorities were left with a number of large and beautifully made henhouses on their hands. There seemed to be no use for them. But suddenly it was discovered that Gambian teachers in training had no proper accommodation. Accordingly, they were put into the henhouses, presumably after some structural adaptation of the buildings.

Clearly it is not so easy to raise production in the tropics as many people seem to think, who glibly say, "Science will find a way." In some cases science may find a way, but it will never do so without a great deal of careful preliminary exploration and a large expenditure of capital, skill, and effort. We must also remember that, even if we do manage to speed up food production so as to keep pace with the extra mouths being born, there is already a backlog of around a billion and a half people who are today undernourished or malnourished, and whose nourishment needs considerable improvement to make it adequate for the development of proper health and strength.

Others suggest cultivating the ocean. Of course it is possible to get more food out of the sea, but here, too, the output is limited; it will not keep up with present rates of population growth for more than a few hundred years at most. Again, it is possible, I am sure, to grow algae and other simple organisms in culture and use them as food. And as Dr. Pirie of Rothamsted has shown, it is possible to extract edible (but apparently not very palatable) protein from the leaves of various plants. There is even a prospect of making complete synthetic food out of simple chemical substances. I am all for trying all such projects. On the other hand, I do not see any reason for pinning our faith entirely on such schemes. Surely it is much less desirable to support ten billion people on unpleasant or synthetic messes than to support half that number on food that is agreeable to eat.

It is also true that irrigation can do a great deal to increase food production. But the benefit is temporary. Irrigation dams silt up, and the irrigated areas soon fill up with people. A classical example is the Lloyd Barrage on the Indus in what is now Pakistan. This was at one time the largest irrigation scheme in the world, and it did make a large area productive. But in a few decades the area became fully settled, at a density almost as high as elsewhere in the region. Another example is the much-discussed High Dam at Aswan in Egypt. This is obviously extremely important, as it will supply the needs of Egypt's growing population for about the next forty years. But, if the rate of population increase does not go down, it will by then have reached the limits of its capacity, and the situation as regards the pressure of population on food resources will be just as it was earlier, except that there will be more people to deal with, and less new land available for irrigation.

Migration is often suggested as a solution to the problem. However, there is not much chance for large-scale migration in the modern world. Everywhere there are strict immigration quotas. Even if Australia threw open its doors to Asian immigrants, the country could not absorb more than fifteen million at the outside; and that represents only one year's increase in China alone! In Indonesia there is the paradoxical situation of the two main islands—overpopulated Java to the east, with extremely high density of people and a high rate of increase, and Sumatra to the west, definitely underpopulated. The Indonesian authorities have quite rightly tried to get people from Java to settle in Sumatra, but have not had much success. There are immense resistances to overcome. The immigrants do not take kindly to the more primitive conditions in Sumatra and tend to migrate back again to Java, or at least to urban settlements on the coast, instead of sticking to a pioneer agricultural life in the interior. The scheme has been very expensive and has resulted merely in the settlement of a few tens of thousands of people, whereas the increase of Java is over a million a year. For Indonesia, migration just does not work.

Sometimes, however, migration is necessary. There are some countries that, without some emigration, will reach an impossible state—they will just burst, so to speak. This applies especially to the West Indian Islands, such as Jamaica or Puerto Rico, where the standard of living and the level of the economy are very low, and the rate of population increase is very high. We are seeing the results of this in the emigration of people from those areas to more prosperous countries. It is comparable to what happens in human cancers, where unlimited multiplication produces what are called metastases—groups of cells that migrate to another part of the body and start trouble there. I would say that it is perfectly legitimate to compare the invasions of Puerto Ricans in New York and of Jamaicans in London and other parts of Britain to metastases in cancer. They are causing considerable social difficulties in their new homes—not primarily because of race prejudice, but because groups of people with different habits and different standards are invading already overcrowded parts of the world.

Indeed, this comparison of overmultiplication with cancer is almost terrifyingly apposite. If man fails to control his rate of increase, he may well cease having any right to call himself the lord of creation and may become the cancer of the whole planet, devouring its resources and exterminating himself, or at least putting paid to his evolutionary hopes.

It has often been suggested that a solution to overpopulation in underdeveloped countries may be found in industrialization. When I had the privilege of an interview with Mr. Nehru before leaving India in 1954, I said that I thought that the government was not pressing on hard enough with their plans for controlling population increase, although these were a part of official policy. He replied in effect that, though population control was important, he felt that quick industrialization was for the moment more essential. It would raise the standard of life, and then they could deal with the population problem. He had forgotten the fact, so elementary that it is often forgotten, that it is impossible to industrialize an underdeveloped country without investing a great deal of capital, not only financial capital, but also human capital in the form of trained manpower and skill. But, if the number of babies born is too great, so much of the financial and human skill capital will be taken up in feeding, educating, housing, and caring for the growing children that not enough will be available for the capital investment needed for industrialization. Soon after this, two distinguished American economists, Ansley J. Coale and Edgar M. Hoover, from Princeton, wrote a report on the relation between industrialization and population increase in India, and they have since written a book on the subject, entitled *Population Growth and Economic Development in Low Income Countries*. As regards India they came to the rather drastic conclusion that, if the country did not manage to cut its birth rate by about 50 per cent in thirty-five or forty years, it would never be able to develop into a viable industrialized nation; it would reach a point of no return, and its standard of life would go down instead of up.

This illustrates the general principle that has to be borne in mind in regard to development projects in underdeveloped countries. This, I feel, is fundamental. All the bodies engaged in giving financial or technical aid and assistance to underdeveloped countries, whether official United Nations organizations like the United Nations itself, the World Bank, the Food and Agricultural Organization, the World Health Organization, or Unesco; or governmental agencies like those implementing the British Colombo plan and the American Association for International Development plan; or private foundations like Ford or Rockefeller, ought to consider it and its implications in relation to all applications for help. When someone applies to a commercial bank for a loan, it is not only legitimate but taken for granted that the bank will look into his creditworthiness. Similarly, the aid-giving agencies ought to look into the demographic creditworthiness of the countries applying for aid or assistance, because if their rate of population increase is too great the financial aid or technical skills provided will be wasted, flushed down the drain of excess population increase. If the nation is judged not to be demographically creditworthy, the aid-giving agency should suggest, tactfully but firmly, that it frame a population policy aimed at bringing down its rate of increase and should assign part of the aid to implementing that policy.

There are other resources by which and on which men live which I can only touch on briefly. As regards metals and other inorganic raw materials, I must simply refer you to books such as Harrison Brown's admirable *The Challenge of Man's Future,* which makes it clear that we are using up these essential raw materials of civilized life at an appalling rate. The process has gone so far that in the United States, the technologically most advanced country in the world, you are having to employ increasingly low-grade ores and increasingly expensive methods of extraction; indeed, in respect to some metals you have had to switch from an exporting to an importing country.

As regards nonrenewable energy resources of coal and oil, there is still quite a reasonable supply left, but it will last only for a few generations, and what is that in the perspective of evolutionary time or even of history? An alarming situation is developing over water. We have been brought up to regard water as virtually inexhaustible, something that we can use to the fullest extent without any scruples. Today, this is no longer so. For instance, in Britain now we are having trouble over the water supply for Manchester. Already in the Lake District, our most beautiful National Park, one of the large lakes has long been utilized for this purpose, and last year the Manchester Corporation wanted to take over the most beautiful of them all, Ullswater, for the same purpose. You on the Northwest Coast may think you are unlucky in having so much rain, but you are really fortunate in having plenty of water. By way of contrast, look at southern California, where Los Angeles has had to fetch water across from the Colorado River and is now talking about bringing it all the seven hundred miles from northern California.

But threats of impending water shortage are beginning to appear all over the world. The world is having to go slow on water, as the demands for irrigation water and industrial water grow ever more insatiable. I am sure that we shall eventually have to resort to distilling sea water. But this will always be expensive; and meanwhile let us harbor the water resources we have and not continue to imagine that they are inexhaustible.

In Britain we are even shorter of space than of water. We are a small island, and there is a real and growing pressure on the space available in it. There is pressure from the Army, the Navy and the Air Force; there are demands for space for atomic power stations, for National Parks, for aerodromes and roads, for industrial development and urban expansion, for forestry and agriculture, and for sport and amenity. Too many interests are competing for too little space. Britain, in fact, is getting overcrowded. The same is true in parts of western Europe, like Holland. The Dutch have done wonders in enlarging their little country by reclaiming land from the sea, but this cannot go on forever.

In the United States you have not yet reached this alarming situation. For one thing, you have a very large country; for another, you are fortunate in having had farsighted men like Theodore Roosevelt and Gifford Pinchot who started your splendid system of National Parks, Forest Reserves, State Parks, and all the rest of it. But, even so, some of your National Parks are beginning to be overcrowded. As a visitor from an overcrowded country, I would urge you to hang on to all the wilderness and recreation space you have; it will become increasingly precious as time goes on. But even in America the increase of population is beginning to press on your resources of space and solitude. In the East you already have an almost continuous belt of human habitation all the way from Boston to Delaware—a belt of urban sprawl, neither town nor country nor suburb, just subtopia. A great deal of the countryside has ceased to be real countryside, and what has grown in its place is not real city.

One of the obvious facts of the modern world is that if the total population of a country increases there will be more big cities in it. An interesting geographical exercise is to look in your encyclopedias and work out for yourself the number of cities over a million that existed in the year 1900 and the corresponding number that exist in this present year. I can assure you that the difference is quite extraordinary.

The rise in the absolute size of the larger cities is also extraordinary. As a young man I traveled a good deal about Europe, and already then, a good half century ago, I came to the conclusion that half a million was the ideal size for a city and that every further increase brought more disadvantages than advantages. I am still of that opinion. Last autumn I spent a week in the pleasant city of Edinburgh and found that it still had less than half a million inhabitants. It has all the doings of civilized life and has easy access to the country. London, on the other hand, is no longer a city but a megalopolis getting on for ten million inhabitants, and life in it is getting impossible, or at least increasingly difficult.

There is a real biological analogy here. It is biologically impossible for evolution to produce a terrestrial animal much bigger than an elephant. It might possibly produce a reasonably efficient animal twice as heavy as an elephant: but an animal four times as heavy as an elephant just would not work. The same thing applies to cities. Cities should be machines for civilized living. As such, they can be made to work up to two million, at a pinch up to three or four million, but further increase makes life in them less and less civilized. It brings increasing congestion, increasing time devoted to daily commuting, increasing traffic problems, until they grind to a halt, seize up, or choke themselves. And yet cities like Tokyo, New York, and London (or rather the conurbations centered on these cities) are now around the ten-million mark and still increasing.

This megalopolitan giantism brings psychological problems, too. The inhabitants of overlarge cities are subjected to an increasing amount of frustration; this in turn is a source of neurosis and inner conflict, which eventually may spill over in fits of irrational aggression. We know that if rats are bred and brought up in overcrowded conditions they get frustrated and emotionally unbalanced, and all kinds of rat social difficulties develop. In man, though the rat race is already with us, the symptoms and effects will not be identical, but they will be similar in principle.

I now come to what may be called enjoyment resources. Man does not exist merely to continue existing. He has other aims in life besides earning his daily bread. He should have at least the possibility of true enjoyment. For the majority of city dwellers—and that means the majority of the population in countries like the United States or Britain—their contact with nature has been cut. I remember a grim story I heard years ago of a little New York boy from the East Side who was taken by his father into the country for the first time. On seeing a rainbow he exclaimed, "Say, Poppa, what is that advertising?"

Things were different in earlier centuries. If you look at Lewis Mumford's remarkable book *The City in History* you will find that right through the Middle Ages, which we affect to despise because they did not have our scientific knowledge and our technological know-how, the inhabitants of even the largest cities still enjoyed many amenities that we lack, notably close contact with nature. On Sundays and holidays they could just walk out of the city into the fields, gather flowers, enjoy themselves in the woods or by a clear river. This was true of London, Rome, Paris—all the great cities of Europe. Today the countryside is being infected by what has been called a bungaloid rash. The remotest beaches are being spoiled by oil, and on some English beaches swimming is no longer possible because of sewage. Natural beauty is being ruined by bad design and by development in the wrong places. The world's wilderness is being invaded; cable railways and ski lifts and teleferics are being pushed up the Swiss Alps and many other noble mountains; and the enjoyment of solitude, which is so important for many people is getting more and more difficult of attainment. Here again you in this country are luckier than your fellow men in Europe; but you must be on your guard, because at your present rate of population growth these precious areas of solitude and enjoyment will be increasingly threatened.

Then there is the problem of education. All human progress depends first on acquiring more and better-organized knowledge and then on transmitting it through more and better education. Today, more and better education is the first demand in all underdeveloped countries, notably perhaps in Africa. But here again population quantity is just as much an obstacle as in other fields. There is a terrific backlog and a terrific deficiency to make up. There are often no textbooks, there are always far too few classrooms, and the supply of teachers is grossly inadequate. Yet all the time more and more children are being born to increase the backlog and enlarge the deficiency. Let me give some examples. There is at the moment a truly explosive demand and an urgent need for better secondary education in all underdeveloped countries; and yet only about 2.5 per cent of the world's child population is getting any secondary education at all. Or take the situation as regards teachers. In my country there are today about 33,000 teachers of all descriptions and levels. By 1970, only eight years hence, a minimum of ninety thousand more will be needed, but nobody knows where they are going to come from. It will be virtually impossible to train a sufficient number of qualified teachers and at the same time to give them sufficiently small classes for their teaching to be effective. And in the United States a reliable survey made in 1956 estimated that it would take more than half the college graduates in the country to supply the teachers needed for the next ten years at the official teacher-pupil ratio. What will happen, of course, is that the official teacher-pupil ratio will not be maintained.

There is the further question of employment. In India today there is widespread white-collar unemployment and an enormous amount of general underemployment—something like seventy million people who are working not more than half time because there just is not enough employment for them. Or take China; even leaving out the whole question of food production, how on earth are the Chinese going to find or make fourteen or fifteen million new jobs every year? Meanwhile all over the world the advance of automation will throw millions of people out of work. Compulsory leisure will be one of the great problems of the future.

Then there is the political aspect. Not only does overcrowding generate a frustrated and quarrelsome spirit, but overpopulation promotes territorial aggression to gain new space. The demand for *Lebensraum* has repeatedly been made the excuse for aggressive war. We saw it in Germany, we saw it in Italy, we saw it in Japan, and I am sure that it was one of the reasons behind China's invasion of Tibet. It will become an increasing threat to peace if population continues to build up at its present rate.

Finally there is the question of social and political efficiency. As population increases beyond a certain point, the need for drastic measures of organization will increase too; more elaborate administrative machinery will be required, more regimentation will be necessary. Thus the increasing pressure of population will inevitably cause overpopulated countries to become increasingly authoritarian in practice or even overtly and deliberately. Overpopulation tends to breed dictatorship.

To sum up, the world's demographic situation is becoming impossible. Man, in the person of the present generation of human beings, is laying a burden on his own future. He is condemning his children's children to increased misery; he is making it harder to improve the general lot of mankind; he is making it more difficult to build a united world free of frustration and greed. More and more human beings will be competing for less and less, or at any rate each will have to be content with a lesser cut of the world's cake. If nothing is done about this problem by us who are now alive, the whole of mankind's future will suffer, including the future of our own children and grandchildren. The next twenty-four years will be decisive.

What ought we to do? The first thing, obviously, is to realize that what is happening is truly a population explosion, that it constitutes the world's most serious problem, and that it has implications in all fields of human life and endeavor. The second is to change the set of our minds, to get away from the idea that increase in the number of human beings is somehow automatically a good thing, something inherently natural and right. It was both natural and right in the early centuries of man's existence and up to fairly recent times, but it is so no longer. Today it is wrong. It is now so wrong that it has become fundamentally immoral for any individual or group or organization to put obstacles in the way of birth control or to oppose any policies aimed at reducing the rate of human increase. In passing, it is an interesting but regrettable fact that the Western world is in this respect less advanced than the Orient. In India or Japan or Pakistan, no important organization is trying to put obstacles in the way of population control; both governments and public have seen that it is necessary and desirable. But here in the West a number of organizations, especially religious organizations, are doing so, often all too successfully.

We must take a new look at the problem. We must stop thinking in terms of a race between production and reproduction, a race that never can be won. We must realize that our aim is not mere quantity, whether of people or goods or anything else, but quality—quality of human beings and of the lives they lead. Once we have grasped this, things begin to fall into place.

Such a view has important implications for science and education. It implies that the most important sciences today for the modern world are not physics and chemistry and their applications in technology, but evolutionary biology and ecology and their applications in scientific conservation.

Ecology is the science of the relations of living things with each other and with their environment. It of course includes human ecology, which deals with the relations of man with his fellow men and with the world's resources, both material resources and psychological or enjoyment resources. So I would plead for much more emphasis in education on biology in general and ecological biology and human ecology in particular.

We must certainly get away from the idea that birth control (or population control, or family planning, or whatever one chooses to call it) is in any way unnatural. It is no more unnatural than wearing clothes or cooking food or traveling in an airplane. Nor is it sinful. It is worth remembering that just over a century ago the use of chloroform for women in childbirth was regarded as wicked and immoral by many churchmen because it was removing the primal curse on woman and not allowing her to pay as she ought to for her original sin in eating the apple and persuading Adam to do the same in the Garden of Eden. These misapplications of theology to the use of anesthetics, to abortion, and now to birth control have caused grievous suffering to an immense number of unfortunate women. It is time they came to an end.

As regards population policy, the time is not ripe for any estimate of ideal or optimum population size for the world or for individual countries. The one overriding aim must be to get the rate of human increase down. Whatever we do, we shall not be achieving an optimum; but we may be able to prevent world disaster if we halve the rate of population increase in the next two or three—preferably two—generations. If we do not manage to reduce the rate of increase, all possibility of future improvement is jeopardized.

This applies, though in different degrees and ways, to all countries and nations. It applies just as much to your nation and my nation as it does to the West Indies or India or China. It will have to be approached in a different way in each nation, since the quantitative factors and the social problems differ from country to country. In the most general terms, we have to aim at establishing a world population policy as part of the broad policy of the United Nations. To implement this, each separate nation should establish its own population policy, with a department of population to implement it.

Meanwhile, of course, we must do everything in our power to increase all types of production—production of food, production of machines, production of what I may call the infrastructure of modern life; but equally of course we must pay the maximum possible attention to the conservation of resources.

That is the aim. How are we to achieve it? One obvious way is to spend much more money and devote much more scientific manpower to research on human reproduction, on methods of birth control and family planning, and on their social implementation in practice. It is fair to say that, if one tenth of one per cent of the money and skilled brain power expended on atomic weapons and space research had gone into research on human reproduction and its control, the world would be in much better shape, and its future would not look so black.

In the field of research it is clear that the scientifically advanced nations, like the United States, Britain, and the countries of western Europe, should give a lead. They are privileged in already possessing highly developed scientific organizations, and they should not only take the lead in doing research in these fields, but should then make the results of that research fully and freely available to the people of all other nations, on request. Internationally as well as nationally, every encouragement should be given to the handful of countries like India, Pakistan, Japan, and some of the West

Indian countries which have already established policies of population control, and they should be aided in every way to implement these policies. And we should encourage other countries to go and do likewise. As I earlier suggested, we should make demographic creditworthiness an element in the granting of aid by the United Nations and all its agencies.

Another thing that the advanced and privileged nations should do is to set their economists and social scientists to thinking out ways and methods of providing economic and social incentives for promoting a lower rate of population increase. Whether by means of family allowances, differential taxation, or other measures, it would undoubtedly be possible to devise economic and social methods that would exert pressure in favor of population decrease. As complement to this, we should set our psychologists and sociologists to studying ways of providing psychological motivation for small families and a sane population policy. In India the authorities are already beginning to try to persuade people that the whole future of the country depends on reducing the birth rate, and consequently that it is unpatriotic to have too many children. This has already been achieved in Japan, with the result that the Japanese have been able to cut their birth rate in half within a generation.

We should support all legislation—state, national, and international—that makes birth control easier and more socially approved. We must start discussion groups and civic action groups and bring pressure to bear on our legislators and our governments. Pressure must also be brought to bear on the United Nations and its agencies. For instance, on two occasions it has been proposed that the World Health Organization should take population density into consideration as a matter affecting the world's health. In both cases the proposal was rejected—so far as I can understand, entirely owing to pressure from Roman Catholic countries and organizations. This is an international scandal. If you can pretend that population pressure does not affect health, you can pretend anything.

There are many other desirable measures we should press for. We should establish official national population policy councils and should organize groups of private citizens who could really be influential in exerting pressure and giving guidance. Then all important universities should have Chairs of Demography. And of course we should see to it that the population problem is adequately publicized. Apropos of this, I did not notice any exhibits on world population at your World's Fair—a sad neglect of the world's most urgent problem. Just remember that during the sixty minutes of my talk the world's population has increased by six thousand human beings. There is indeed a population explosion in progress.

Meanwhile, it is fair to say that the facts of the population explosion have begun to make people ask the most fundamental of all questions, so fundamental that we usually do not bother to ask it: "What are people *for*?" This brings me back to my earlier conclusion that our overriding objective must be richer fulfillment for more individual human beings and fuller achievement by more human communities. If you prefer it, man's ultimate destiny is to direct the process of evolution and lead it to new heights by realizing new possibilities to enhance the quality of human living.

Today, our knowledge is bringing out ever more clearly the extent of human possibilities which are now unrealized. I believe that once we grasp the extent and the wonder of these unrealized possibilities they will provide the great motivation for concerted action. The prospect of exploiting the vast treasure of latent potentialities to a point where the average man would be realizing possibilities that today are realized only in a handful of favored individuals—this, I believe, might provide the great motive and mainspring of man's future evolutionary course. Here universities and scientific organizations can play a part. As I said in my first lecture, Margaret Mead has proposed that universities should establish "Chairs of the Future," and I suggested that we should recognize the science of human possibilities as one of the major fields of scientific investigation. Such measures would not only help us to find the right dominant motivation for future joint action by mankind, but would also provide compass bearings for the direction of our future advance in this unexplored but rich territory of existence.

Sir Julian Huxley and Kingsley Davis propose voluntary and socioeconomic approaches to the population problem. Garrett Hardin (1915-) carries the discussion one step further: As long as there exists the unlimited freedom to breed, Hardin claims, there will be no solution to the population problem. Comparing the earth to a commons, a pasture for the use of all, Hardin reminds us that freedom in a commons brings ruin to all. Freedom to exploit the resources of the commons, freedom to pollute the commons, freedom to overpopulate the commons—all of these are intolerable to those who depend upon the commons for their existence. We must, says Hardin, turn to laws rather than to appeals to conscience or responsibility to stop this destruction of the commons. Most importantly, the freedom to breed must be abandoned in order to preserve more important freedoms. Zero population growth must soon be achieved in this finite world.

Garrett Hardin is a professor of biology at the University of California at Santa Barbara and a well-known scholar in the fields of population and human ecology.

POPULATION POLICY: WILL CURRENT PROGRAMS SUCCEED?

by Kingsley Davis

Throughout history the growth of population has been identified with prosperity and strength. If today an increasing number of nations are seeking to curb rapid population growth by reducing their birth rates, they must be driven to do so by an urgent crisis. My purpose here is not to discuss the crisis itself but rather to assess the present and prospective measures used to meet it. Most observers are surprised by the swiftness with which concern over the population problem has turned from intellectual analysis and debate to policy and action. Such action is a welcome relief from the long opposition, or timidity, which seemed to block forever any governmental attempt to restrain population growth, but relief that "at last something is being done" is no guarantee that what is being done is adequate. On the face of it, one could hardly expect such a fundamental re-orientation to be quickly and successfully implemented. I therefore propose to review the nature and (as I see them)

"Population Policy: Will Current Programs Succeed?" From *Science*, November 10, 1967, Vol. 158, pp. 730-739.

limitations of the present policies and to suggest lines of possible improvement.

With more than 30 nations now trying or planning to reduce population growth and with numerous private and international organizations helping, the degree of unanimity as to the kind of measures needed is impressive. The consensus can be summed up in the phrase "family planning." President Johnson declared in 1965 that the United States will "assist family planning programs in nations which request such help." The Prime Minister of India said a year later, "We must press forward with family planning. This is a programme of the highest importance." The Republic of Singapore created in 1966 the Singapore Family Planning and Population Board "to initiate and undertake population control programmes."[1]

As is well known, "family planning" is a euphemism for contraception. The family-planning approach to population limitation, therefore, concentrates on providing new and efficient contraceptives on a national basis through mass programs under public health auspices. The nature of these programs is shown by the following enthusiastic report from the Population Council:

> *No single year has seen so many forward steps in population control as 1965. Effective national programs have at last emerged, international organizations have decided to become engaged, a new contraceptive has proved its value in mass application, . . . and surveys have confirmed a popular desire for family limitation . . .*
>
> *An accounting of notable events must begin with Korea and Taiwan . . . Taiwan's program is not yet two years old, and already it has inserted one IUD [intrauterine device] for every 4-6 target women (those who are not pregnant, lactating, already sterile, already using contraceptives effectively, or desirous of more children). Korea has done almost as well . . . has put 2,200 full-time workers into the field, . . . has reached operational levels for a network of IUD quotas, supply lines, local manufacture of contraceptives, training of hundreds of M.D.'s and nurses, and mass propaganda . . .*[2]

Here one can see the implication that "population control" is being achieved through the dissemination of new contraceptives, and the fact that the "target women" exclude those who want more children. One can also note the technological emphasis and the medical orientation.

What is wrong with such programs? The answer is, "Nothing at all, if they work." Whether or not they work depends on what they are expected to do as well as on how they try to do it. Let us discuss the goal first, then the means.

Curiously, it is hard to find in the population-policy movement any explicit discussion of long-range goals. By implication the policies seem to promise a great deal. This is shown by the use of expressions like *population control* and *population planning* (as in the passages quoted above). It is also shown by the characteristic style of reasoning. Expositions of current policy usually start off by lamenting the speed and the consequences of runaway population growth. This growth, it is then stated, must be curbed—by pursuing a vigorous family-planning program. That family planning can solve the problem of population growth seems to be taken as self-evident.

For instance, the much-heralded statement by 12 heads of state, issued by Secretary-General U Thant on 10 December 1966 (a statement initiated by John D. Rockefeller III, Chairman of the Board of the Population Council), devotes half its space to discussing the harmfulness of population growth and the other half to recommending family planning.[3] A more succinct example of the typical reasoning is given in the Provisional Scheme for a Nationwide Family Planning Programme in Ceylon: "The Population of Ceylon is fast increasing. . . . [The] figures reveal that a serious situation will be created within a few years. In order to cope with it a Family Planning programme on a nationwide scale should be launched by the Government."[4] The promised goal—to limit population growth so as to solve population problems—is a large order. One would expect it to be carefully analyzed, but it is left imprecise and taken for granted, as is the way in which family planning will achieve it.

When the terms *population control* and *population planning* are used, as they frequently are, as synonyms for current family-planning programs, they are misleading. Technically, they would mean deliberate influence over all attributes of a population, including its age-sex structure, geographical distribution, racial composition, genetic quality, and total size. No government attempts such full control. By tacit understanding, current population policies are concerned with only the *growth* and *size* of populations. These attributes, however, result from the death rate and migration as well as from the birth rate; their control would require deliberate influence over the factors giving rise to all three determinants. Actually, current policies labeled population control do not deal with mortality and migration, but deal only with the birth input. This is why another term, *fertility control,* is frequently used to describe current policies. But, as I show below, family planning (and hence current policy) does not undertake to influence most of the determinants of human reproduction. Thus the programs should not be referred to as population control or planning, because they do not attempt to influence the factors responsible for the attributes of human populations, taken generally; nor should they be called fertility control, because they do not try to affect most of the determinants of reproductive performance.

The ambiguity does not stop here, however. When one speaks of controlling population size, any inquiring person naturally asks, What is "control"? Who is to control whom? Precisely what population size, or what rate of population growth is to be achieved? Do the policies aim to produce a growth rate that is nil, one that is very slight, or one that is like that of the industrial nations? Unless such questions are dealt with and clarified, it is impossible to evaluate current population policies.

The actual programs seem to be aiming simply to achieve a reduction in the birth rate. Success is therefore interpreted as the accomplishment of such a reduction, on the assumption that the reduction will lessen population growth. In those rare cases where a specific demographic aim is stated, the goal is said to be a short-run decline within a given period. The Pakistan plan adopted in 1966[5] aims to reduce the birth rate from 50 to 40 per thousand by 1970; the Indian plan[6] aims to reduce the rate from 40 to 25 "as soon as possible"; and the Korean aim[7] is to cut population growth from 2.9 to 1.2 percent by 1980. A significant feature of such stated aims is the rapid population growth they would permit. Under conditions of modern mortality, a crude birth rate of 25 to 30 per thousand will represent such a multiplication of people as to make use of the term *population control* ironic. A rate of increase of 1.2 percent per year would allow South Korea's already dense population to double in less than 60 years.

One can of course defend the programs by saying that the present goals and measures are merely interim ones. A start must be made somewhere. But we do not find this answer in the population-policy literature. Such a defense, if convincing, would require a presentation of the *next* steps, and these are not considered. One suspects that the entire question of goals is instinctively left vague because thorough limitation of population growth would run counter to national and group aspirations. A consideration of hypothetical goals throws further light on the matter.

Industrialized nations as the model. Since current policies are confined to family planning, their maximum demographic effect would be to give the underdeveloped countries the same level of reproductive performance that the industrial nations now have. The latter, long oriented toward family planning, provide a good yardstick for determining what the availability of contraceptives can do to population growth. Indeed, they provide more than a yardstick; they are actually the model which inspired the present population policies.

What does this goal mean in practice? Among the advanced nations there is considerable diversity in the level of fertility.[8] At one extreme are countries such as New Zealand, with an average gross reproduction rate (GRR) of 1.91 during the period 1960-64; at the other extreme are countries such as Hungary, with a rate of 0.91 during the same period. To a considerable extent, however, such divergencies are matters of timing. The birth rates of most industrial nations have shown, since about 1940, a wave-like movement, with no secular trend. The average level of reproduction during this long period has been high enough to give these countries, with their low mortality, an extremely rapid population growth. If this level is maintained, their population will double in just over 50 years—a rate higher than that of world population growth at any time prior to 1950, at which time the growth in numbers of human beings was already considered fantastic. The advanced nations are suffering acutely from the effects of rapid population growth in combination with the production of ever more goods per person.[9] A rising share of their supposedly high per capita income, which itself draws increasingly upon the resources of the underdeveloped countries (who fall farther behind in relative economic position), is spent simply to meet the costs, and alleviate the nuisances, of the unrelenting production of more and more goods by more people. Such facts indicate that the industrial nations provide neither a suitable demographic model for the nonindustrial peoples to follow nor the leadership to plan and organize effective population-control policies for them.

Zero population growth as a goal. Most discussions of the population crisis lead logically to zero population growth as the ultimate goal, because *any* growth rate, if continued, will eventually use up the earth. Yet hardly ever do arguments for population policy consider such a goal, and current policies do not dream of it. Why not? The answer is evidently that zero population growth is unacceptable to most nations and to most religious and ethnic communities. To argue for this goal would be to alienate possible support for action programs.

Goal peculiarities inherent in family planning. Turning to the actual measures taken, we see that the very use of family planning as the means for implementing population policy poses serious but unacknowledged limits on the intended reduction in fertility. The family-planning movement, clearly devoted to the improvement and dissemination of contraceptive devices, states again and again that its purpose is that of enabling couples to have the number of children they want. "The opportunity to decide the number and spacing of children is a basic human right," say the 12 heads of state in the United Nations declaration. The 1965 Turkish Law Concerning Population Planning declares: "*Article 1.* Population Planning means that individuals can have as many children as they wish, whenever they want to. This can be ensured through preventive measures taken against pregnancy. . . ."[10]

Logically, it does not make sense to use *family* planning to provide *national* population control or planning. The "planning" in family planning is that of each separate couple. The only control they exercise is control over the size of *their* family. Obviously, couples do not plan the size of the nation's population, any more than they plan the growth of the national income or the form of the highway network. There is no reason to expect that the millions of decisions about family size made by couples in their own interest will automatically control population for the benefit of society. On the contrary, there are good reasons to think they will not do so. At most, family planning can reduce reproduction to the extent that unwanted births exceed wanted births. In industrial countries the balance is often negative—that is, people have fewer children as a rule than they would like to have. In underdeveloped countries the reverse is normally true, but the elimination of unwanted births would still leave an extremely high rate of multiplication.

Actually, the family-planning movement does not pursue even the limited goals it professes. It does not fully empower couples to have only the number of offspring they want because it either condemns or disregards certain tabooed but nevertheless effective means to this goal. One of its tenets is that "there shall be freedom of choice of method so that individuals can choose in accordance with the dictates of their consciences,"[11] but in practice this amounts to limiting the individual's choice, because the "conscience" dictating the method is usually not his but that of religious and governmental officials. Moreover, not every individual may choose: even the so-called recommended methods are ordinarily not offered to single women, or not all offered to women professing a given religious faith.

Thus, despite its emphasis on technology, current policy does not utilize all available means of contraception, much less all birth-control measures. The Indian government wasted valuable years in the early stages of its population-control program by experimenting exclusively with the "rhythm" method, long after this technique had been demonstrated to be one of the least effective. A greater limitation on means is the exclusive emphasis on contraception itself. Induced abortion, for example, is one of the surest means of controlling reproduction, and one that has been proved capable of reducing birth rates rapidly. It seems peculiarly suited to the threshold stage of a population-control program—the stage when new conditions of life first make large families disadvantageous. It was the principal factor in the halving of the Japanese birth rate, a major factor in the declines in birth rate of East-European satellite countries after legalization of abortions in the early 1950's, and an important factor in the reduction of fertility in industrializing nations from 1870 to the 1930's.[12] Today, according to *Studies in Family Planning,* "abortion is probably the foremost method of birth control throughout Latin America."[13] Yet this method is rejected in nearly all national and international population-control programs. American foreign aid is used to help *stop* abortion.[14] The United Nations excludes abortion from family planning, and in fact justifies the latter by presenting it as a means of combating abortion.[15] Studies of abortion are being made in Latin America under the presumed auspices of population-control groups, not with the intention of legalizing it and thus making it safe, cheap, available, and hence more effective for population control, but with the avowed purpose of reducing it.[16]

Although few would prefer abortion to efficient contraception (other things being equal), the fact is that both permit a woman to control the size of her family. The main drawbacks to abortion arise from its illegality. When performed, as a legal procedure, by a skilled physician, it is safer than childbirth. It does not compete with contraception but serves as a backstop when the latter fails or when contraceptive devices or information are not available. As contraception becomes customary, the incidence of abortion recedes even without its being banned, If, therefore, abortions enable women to have only the number of children they want, and if family planners do not advocate—in fact decry—legalization of abortion, they are to that extent denying the central tenet of their own movement. The irony of anti-abortionism in family-planning circles is seen particularly in hair-splitting arguments over whether or not some contraceptive agent (for example, the IUD) is in reality an abortifacient. A Mexican leader in family planning writes: "One of the chief objectives of our program in Mexico is to prevent abortions. If we could be sure that the mode of action [of the IUD] was not interference with nidation, we could easily use the method in Mexico."[17]

The questions of sterilization and unnatural forms of sexual intercourse usually meet with similar silent treatment or disapproval, although nobody doubts the effectiveness of these measures in avoiding conception. Sterilization has proved popular in Puerto Rico and has had some vogue in India (where the new health minister hopes to make it compulsory for those with a certain number of children), but in both these areas it has been for the most part ignored or condemned by the family-planning movement.

On the side of goals, then, we see that a family-planning orientation limits the aims of current population policy. Despite reference to "population control" and "fertility control," which presumably mean determination of demographic results by and for the nation as a whole, the movement gives control only to couples, and does this only if they use "respectable" contraceptives.

By sanctifying the doctrine that each woman should have the number of children she wants, and by assuming that if she has only that number this will automatically curb population growth to the necessary degree, the leaders of current policies escape the necessity of asking why women desire so many children and how this desire can be influenced.[18,19] Instead, they claim that satisfactory motivation is shown by the popular desire (shown by opinion surveys in all countries) to have the means of family limitation, and that therefore the problem is one of inventing and distributing the best possible contraceptive devices. Overlooked is the fact that a desire for availability of contraceptives is compatible with *high* fertility.

Given the best of means, there remain the questions of how many children couples want and of whether this is the requisite number from the standpoint of population size. That it is not is indicated by continued rapid population growth in industrial countries, and by the very surveys showing that people want contraception—for these show, too, that people also want numerous children.

The family planners do not ignore motivation. They are forever talking about "attitudes" and "needs." But they pose the issue in terms of the "acceptance" of birth control devices. At the most naive level, they assume that lack of acceptance is a function of the contraceptive device itself. This reduces the motive problem to a technological question. The task of population control then becomes simply the invention of a device that *will* be acceptable.[20] The plastic IUD is acclaimed because, once in place, it does not depend on repeated *acceptance* by the woman, and thus it "solves" the problem of motivation.[21]

But suppose a woman does not want to use *any* contraceptive until after she has had four children. This is the type of question that is seldom raised in the family-planning literature. In that literature, wanting a specific number of children is taken as complete motivation, for it implies a wish to control the size of one's family. The problem woman, from the standpoint of family planners, is the one who wants "as many as come," or "as many as God sends." Her attitude is construed as due to ignorance and "cultural values," and the policy deemed necessary to change it is "education." No compulsion can be used, because the movement is committed to free choice, but movie strips, posters, comic books, public lectures, interviews, and discussions are in order. These supply information and supposedly change values by discounting superstitions and showing that unrestrained procreation is harmful to both mother and children. The effort is considered successful when the woman decides she wants only a certain number of children and uses an effective contraceptive.

In viewing negative attitudes toward birth control as due to ignorance, apathy, and outworn tradition, and "mass-communication" as the solution to the motivation problem,[22] family planners tend to ignore the power and complexity of social life. If it were admitted that the creation and care of new human beings is socially motivated, like other forms of behavior, by being a part of the system of rewards and punishments that is built into human relationships, and thus is bound up with the individual's economic and personal interests, it would be apparent that the social structure and economy must be changed before a deliberate reduction in the birth rate can be achieved. As it is, reliance on family planning allows people to feel that "something is being done about the population problem" without the need for painful social changes.

Designation of population control as a medical or public health task leads to a similar evasion. This categorization assures popular support because it puts population policy in the hands of respected medical personnel, but, by the same token, it gives responsibility for leadership to people who think in terms of clinics and patients, of pills and IUD's, and who bring to the handling of economic and social phenomena a self-confident naivete. The study of social organization is a technical field; an action program based on intuition is no more apt to succeed in the control of human beings than it is in the area of bacterial or viral control. Moreover, to alter a social system, by deliberate policy, so as to regulate births in accord with the demands of the collective welfare would require political power, and this is not likely to inhere in public health officials, nurses, midwives, and social workers. To entrust population policy to them is "to take action," but not dangerous "effective action."

Similarly, the Janus-faced position on birth-control technology represents an escape from the necessity, and onus, of grappling with the social and economic determinants of reproductive behavior. On the one side, the rejection or avoidance of religiously tabooed but otherwise effective means of birth prevention enables the family-planning movement to avoid official condemnation. On the other side, an intense preoccupation with contraceptive technology (apart from the tabooed means) also helps the family planners to avoid censure. By implying that the only need is the invention and distribution of effective contraceptive devices, they allay fears, on the part of religious and governmental officials, that fundamental changes in social organization are contemplated. Changes basic enough to affect motivation for having children would be changes in the structure of the family, in the position of women, and in the sexual mores. Far from proposing such radicalism, spokesmen for family planning frequently state their purpose as "protection" of the family—that is, closer observance of family norms. In addition, by concentrating on *new* and *scientific* contraceptives, the movement escapes taboos attached to old ones (the Pope will hardly authorize the condom, but may sanction the pill) and allows family planning to be regarded as a branch of medicine: overpopulation becomes a disease, to be treated by a pill or a coil.

We thus see that the inadequacy of current population policies with respect to motivation is inherent in their overwhelmingly family-planning character. Since family planning is by definition private planning, it eschews any societal control over motivation. It merely furnishes the means, and, among possible means, only the most respectable. Its leaders, in avoiding social complexities and seeking official favor, are obviously activated not solely by expediency but also by their own sentiments as members of society and by their background as persons attracted to the family-planning movement. Unacquainted for the most part with technical economics, sociology, and demography, they tend honestly and instinctively to believe that something they vaguely call population control can be achieved by making better contraceptives available.

If this characterization is accurate, we can conclude that current programs will not enable a government to control population size. In countries where couples have numerous offspring that they do not want, such programs may possibly accelerate a birth-rate decline that would occur anyway, but the conditions that cause births to be wanted or unwanted are beyond the control of any nation which relies on family planning alone as its population policy.

This conclusion is confirmed by demographic facts. As I have noted above, the widespread use of family planning in industrial countries has not given their governments control over the birth rate. In backward countries today, taken as a whole, birth rates are rising, not falling; in those with population policies, there is no indication that the government is controlling the rate of reproduction. The main "successes" cited in the well-publicized policy literature are cases where a large number of contraceptives have been distributed or where the program has been accompanied by some decline in the birth rate. Popular enthusiasm for family planning is found mainly in the cities, or in advanced countries such as Japan and Taiwan, where the people would adopt contraception in any case, program or no program. It is difficult to prove that present population policies have even speeded up a lowering of the birth rate (the least that could have been expected), much less that they have provided national "fertility control."

Let us next briefly review the facts concerning the level and trend of population in underdeveloped nations generally, in order to understand the magnitude of the task of genuine control.

In ten Latin-American countries, between 1940 and 1959,[23] the average birth rates (age-standardized), as estimated by our research office at the University of California, rose as follows: 1940-44, 43.4 annual births per 1000 population; 1945-49, 44.6; 1950-54, 46.4; 1955-59, 47.7.

In another study made in our office, in which estimating methods derived from the theory of quasi-stable populations were used, the recent trend was found to be upward in 27 underdeveloped countries, downward in six, and unchanged in one.[24] Some of the rises have been substantial, and most have occurred where the birth rate was already extremely high. For instance, the gross reproduction rate rose in Jamaica from 1.8 per thousand in 1947 to 2.7 in 1960; among the natives of Fiji, from 2.0 in 1951 to 2.4 in 1964; and in Albania, from 3.0 in the period 1950-54 to 3.4 in 1960.

The general rise in fertility in backward regions is evidently not due to failure of population-control efforts, because most of the countries either have no such effort or have programs too new to show much effect. Instead, the rise is due, ironically, to the very circumstance that brought on the population crisis in the first place—to improved health and lowered mortality. Better health increases the probability that a woman will conceive and retain the fetus to term; lowered mortality raises the proportion of babies who survive to the age of reproduction and reduces the probability of widowhood during that age.[25] The significance of the general rise in fertility, in the context of this discussion, is that it is giving would-be population planners a harder task than many of them realize. Some of the upward pressure on birth rates is independent of what couples do about family planning, for it arises from the fact that, with lowered mortality, there are simply more couples.

In discussions of population policy there is often confusion as to which cases are relevant. Japan, for instance, has been widely praised for the effectiveness of its measures, but it is a very advanced industrial nation and, besides, its government policy had little or nothing to do with the decline in the birth rate, except unintentionally. It therefore offers no test of population policy under peasant-agrarian conditions. Another case of questionable relevance is that of Taiwan, because Taiwan is sufficiently developed to be placed in the urban-industrial class of nations. However, since Taiwan is offered as the main showpiece by the sponsors of current policies in underdeveloped areas, and since the data are excellent, it merits examination.

Taiwan is acclaimed as a showpiece because it has responded favorably to a highly organized program for distributing up-to-date contraceptives and has also had a rapidly dropping birth rate. Some observers have carelessly attributed the decline in the birth rate—from 50.0 in 1951 to 32.7 in 1965—to the family-planning campaign,[26] but the campaign began only in 1963 and could have affected only the end of the trend. Rather, the decline represents a response to modernization similar to that made by all countries that have become industrialized.[27] By 1950 over half of Taiwan's population was urban, and by 1964 nearly two-thirds were urban, with 29 percent of the population living in cities of 100,000 or more. The pace of economic development has been extremely rapid. Between 1951 and

DECLINE IN TAIWAN'S FERTILITY RATE, 1951-1966 *

Year	Registered Births per 1,000 women, aged 15-49	Change in rate (percent)
1951	211	
1952	198	−5.6
1953	194	−2.2
1954	193	−0.5
1955	197	+2.1
1956	196	−0.4
1957	182	−7.1
1958	185	+1.3
1959	184	−0.1
1960	180	−2.5
1961	177	−1.5
1962	174	−1.5
1963	170	−2.6
1964	162	−4.9
1965	152	−6.0
1966	149	−2.1

* The percentages were calculated on unrounded figures. Source of data through 1965, *Taiwan* Demographic Fact Book (1964, 1965); for 1966, *Monthly Bulletin of Population Registration Statistics of Taiwan* (1966, 1967).

BIRTHS PER 1,000 WOMEN AGED 15 THROUGH 49

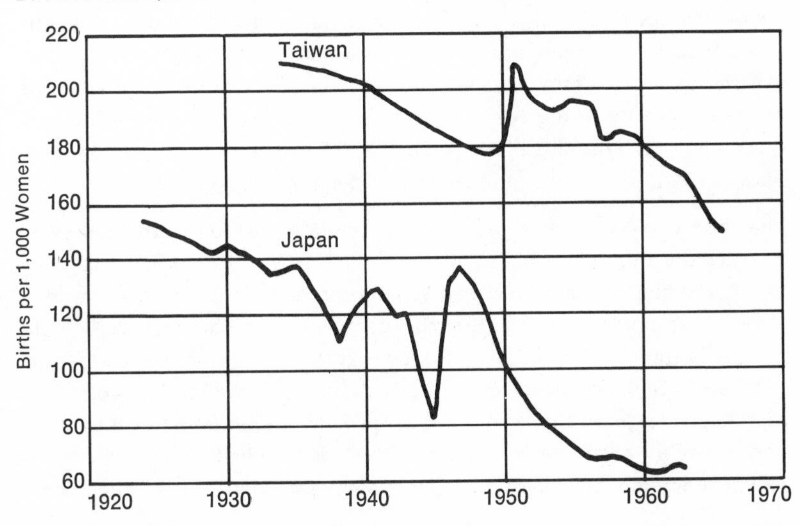

1963, per capita income increased by 4.05 percent per year. Yet the island is closely packed, having 870 persons per square mile (a population density higher than that of Belgium). The combination of fast economic growth and rapid population increase in limited space has put parents of large families at a relative disadvantage and has created a brisk demand for abortions and contraceptives. Thus the favorable response to the current campaign to encourage use of the IUD is not a good example of what birth-control technology can do for a genuinely backward country. In fact, when the program was started, one reason for expected receptivity was that the island was already on its way to modernization and family planning.[28]

At most, the recent family-planning campaign—which reached significant proportions only in 1964, when some 46,000 IUD's were inserted (in 1965 the number was 99,253, and in 1966, 111,242)[29,30] —could have caused the increase observable after 1963 in the rate of decline. Between 1951 and 1963 the average drop in the birth rate per 1000 women (see Table 1) was 1.73 percent per year; in the period 1964-66 it was 4.35 percent. But one hesitates to assign all of the acceleration in decline since 1963 to the family-planning campaign. The rapid economic development has been precisely of a type likely to accelerate a drop in reproduction. The rise in manufacturing has been much greater than the rise in either agriculture or construction. The agricultural labor force has thus been squeezed, and migration to the cities has skyrocketed.[31] Since housing has not kept pace, urban families have had to restrict reproduction in order to take advantage of career opportunities and avoid domestic inconvenience. Such conditions have historically tended to accelerate a decline in birth rate. The most rapid decline came late in the United States (1921-33) and in Japan (1947-55). A plot of the Japanese and Taiwanese birth rates (Fig. 1) shows marked similarity of the two curves, despite a difference in level. All told, one should not attribute all of the post-1963 acceleration in the decline of Taiwan's birth rate to the family-planning campaign.

The main evidence that *some* of this acceleration is due to the campaign comes from the fact that Taichung, the city in which the family-planning effort was first concentrated, showed subsequently a much faster drop in fertility than other cities.[32,33] But the campaign has not reached throughout the island. By the end of 1966, only 260,745 women had been fitted with an IUD under auspices of the campaign, whereas the women of reproductive age on the island numbered 2.86 million. Most of the reduction in fertility has therefore been a matter of individual initiative. To some extent the campaign may be simply substituting sponsored (and cheaper) services for those that would otherwise come through private and commercial channels. An island-wide survey in 1964 showed that over 150,000 women were already using the traditional Ota ring (a metallic intrauterine device popular in Japan); almost as many had been sterilized; about 40,000 were using foam tablets; some 50,000 admitted to having had at least one abortion; and many were using other methods of birth control.[34]

The important question, however, is not whether the present campaign is somewhat hastening the downward trend in the birth rate but whether, even if it is, it will provide population control for the nation. Actually, the campaign is not designed to provide such control and shows no sign of doing so. It takes for granted existing reproductive goals. Its aim is "to integrate, through education and information, the idea of family limitation *within the existing attitudes, values and goals* of the people. [italics mine].[35] Its target is *married* women who do not want any more children; it ignores girls not yet married, and women married and wanting more children.

With such an approach, what is the maximum impact possible? It is the difference between the number of children women have been having and the number they want to have. A study in 1957 found a median figure of 3.75 for the number of children wanted by women aged 15 to 29 in Taipei, Taiwan's largest city; the corresponding figure for women from a satellite town was 3.93; for women from a fishing village, 4.90; and for women from a farming village, 5.03. Over 60 percent of the women in Taipei and over 90 percent of those in the farming village wanted 4 or more children.[36] In a sample of wives aged 25 to 29 in Taichung, a city of over 300,000, Freedman and his co-workers found the average number of children wanted was 4; only 9 percent wanted less than 3, 20 percent wanted 5 or more.[37] If, therefore, Taiwanese women used contraceptives that were 100-percent effective and had the number of children they desire, they would have about 4.5 each. The goal of the family-planning effort would be achieved. In the past the Taiwanese woman who married and lived through the reproductive period had, on the average, approximately 6.5 children; thus a figure of 4.5 would represent a substantial decline in fertility. Since mortality would continue to decline, the population growth rate would decline somewhat less than individual reproduction would. With 4.5 births per woman and a life expectancy of 70 years, the rate of natural increase would be close to 3 percent per year.[38]

In the future, Taiwanese views concerning reproduction will doubtless change, in response to social change and economic modernization. But how far will they change? A good indication is the number of children desired by couples in an already modernized country long oriented toward family planning. In the United States in 1966, an average of 3.4 children was considered ideal by white women aged 21 or over.[39] This average number of births would give Taiwan, with only a slight decrease in mortality, a long-run rate of natural increase of 1.7 percent per year and a doubling of population in 41 years.

Detailed data confirm the interpretation that Taiwanese women are in the process of shifting from a "peasant-agrarian" to an "industrial" level of reproduction. They are, in typical fashion, cutting off higher-order births at age 30 and beyond.[40] Among young wives, fertility has risen, not fallen. In sum, the widely acclaimed family-planning program in Taiwan may, at most, have somewhat speeded the later phase of fertility decline which would have occurred anyway because of modernization.

Moving down the scale of modernization, to countries most in need of population control, one finds the family-planning approach even more inadequate. In South Korea, second only to Taiwan in the frequency with which it is cited as a model of current policy, a recent birth-rate decline of unknown extent is assumed by leaders to be due overwhelmingly to the government's family-planning program. However, it is just as plausible to say that the net effect of government involvement in population control has been, so far, to delay rather than hasten a decline in reproduction made inevitable by social and economic changes. Although the government is advocating vasectomies and providing IUD's and pills, it refuses to legalize abortions, despite the rapid rise in the rate of illegal abortions and despite the fact that, in a recent survey, 72 percent of the people who stated an opinion favored legalization. Also, the program is presented in the context of maternal and child health; it thus emphasizes motherhood and the family rather than alternative roles for women. Much is made of the fact that opinion surveys show as overwhelming majority of Koreans (89 percent in 1965) favoring contraception,[41] but this means only that Koreans are like other people in wishing to have the means to get what they want. Unfortunately, they want sizable families: "The records indicate that the program appeals mainly to women in the 30-39 year age bracket who have four or more children, including at least two sons . . ."[42]

In areas less developed than Korea the degree of acceptance of contraception tends to be disappointing, especially among the rural majority. Faced with this discouragement, the leaders of current policy, instead of reexamining their assumptions, tend to redouble their effort to find a contraceptive that will appeal to the most illiterate peasant, forgetting that he wants a good-sized family. In the rural Punjab, for example, "a disturbing feature . . . is that the females start to seek advice and adopt family planning techniques at the fag end of their reproductive period."[43] Among 5196 women coming to rural Punjabi family-planning centers, 38 percent were over 35 years old, 67 percent over 30. These women had married early, nearly a third of them before the age of 15;[44] some 14 percent had eight or more *living* children when they reached the clinic, 51 percent six or more.

A survey in Tunisia showed that 68 percent of the married couples were willing to use birth-control measures, but the average number of children they considered ideal was 4.3.[45] The corresponding averages for a village in eastern Java, a village near New Delhi, and a village in Mysore were 4.3, 4.0, and 4.2, respectively.[46,47] In the cities of these regions women are more ready to accept birth control and they want fewer children than village women do, but the number they consider desirable is still wholly unsatisfactory from the standpoint of population control. In an urban family-planning center in Tunisia, more than 600 of 900 women accepting contraceptives had four living children already.[48] In Bangalore, a city of nearly a million at the time (1952), the number of offspring desired by married women was 3.7 on the average; by married men, 4.1.[49] In the metropolitan area of San Salvador (350,000 inhabitants) a 1964 survey[50] showed the number desired by women of reproductive age to be 3.9, and in seven other capital cities of Latin America the number ranged from 2.7 to 4.2. If women in the cities of underdeveloped countries used birth-control measures with 100-percent efficiency, they still would have enough babies to expand city populations senselessly, quite apart from the added contribution of rural-urban migration. In many of the cities the difference between actual and ideal number of children is not great; for instance, in the seven Latin-American capitals mentioned above, the ideal was 3.4 whereas the actual births per women in the age range 35 to 39 was 3.7.[51] Bombay City has had birth-control clinics for many years, yet its birth rate (standardized for age, sex, and marital distribution) is still 34 per 1000 inhabitants and is tending to rise rather than fall. Although this rate is about 13 percent lower than that for India generally, it has been about that much lower since at least 1951.[52]

To acknowledge that family planning does not achieve population control is not to impugn its value for other purposes. Freeing women from the need to have more children than they want is of great benefit to them and their children and to society at large. My argument is therefore directed not against family-planning programs as such but against the assumption that they are an effective means of controlling population growth.

But what difference does it make? Why not go along for awhile with family planning as an initial approach to the problem of population control? The answer is that any policy on which millions of dollars are being spent should be designed to achieve the goal it purports to achieve. If it is only a first step, it should be so labeled, and its connection with the next step (and the nature of that next step) should be carefully examined. In the present case, since no "next step" seems ever to be mentioned, the question arises, Is reliance on family planning in fact a basis for dangerous postponement of effective steps? To continue to offer a remedy as a cure long after it has been shown merely to ameliorate the disease is either quackery or wishful thinking, and it thrives most where the need is greatest. Today the desire to solve the population problem is so intense that we are all ready to embrace any "action program" that promises relief. But postponement of effective measures allows the situation to worsen.

Unfortunately, the issue is confused by a matter of semantics. "Family *planning*" and "fertility *control*" suggest that reproduction is being regulated according to some rational plan. And so it is, but only from the standpoint of the individual couple, not from that of the community. What is rational in the light of a couple's situation may be totally irrational from the standpoint of society's welfare.

The need for societal regulation of individual behavior is readily recognized in other spheres—those of explosives, dangerous drugs, public property, natural resources. But in the sphere of reproduction, complete individual initiative is generally favored even by those liberal intellectuals who, in other spheres, most favor economic and social planning. Social reformers who would not hesitate to force all owners of rental property to rent to anyone who can pay, or to force all workers in an industry to join a union, balk at any suggestion that couples be permitted to have only a certain number of offspring. Invariably they interpret societal control of reproduction as meaning direct police supervision of individual behavior. Put the word *compulsory* in front of any term describing a means of limiting births —*compulsory sterilization, compulsory abortion, compulsory contraception* —and you guarantee violent opposition. Fortunately, such direct controls need not be invoked, but conservatives and radicals alike overlook this in their blind opposition to the idea of collective determination of a society's birth rate.

That the exclusive emphasis on family planning in current population policies is not a "first step" but an escape from the real issues is suggested by two facts. (i) No country has taken the "next step." The industrialized countries have had family planning for half a century without acquiring control over either the birth rate or population increase. (ii) Support and encouragement of research on population policy other than family planning is negligible. It is precisely this blocking of alternative thinking and experimentation that makes the emphasis on family planning a major obstacle to population control. The need is not to abandon family-planning programs but to put equal or greater resources into other approaches.

In thinking about other approaches, one can start with known facts. In the past, all surviving societies had institutional incentives for marriage, procreation, and child care which were powerful enough to keep the birth rate equal to or in excess of a high death rate. Despite the drop in death rates during the last century and a half, the incentives tended to remain intact because the social structure (especially in regard to the family) changed little. At most, particularly in industrial societies, children became less productive and more expensive.[53] In present-day agrarian societies, where the drop in death rate has been more recent, precipitate, and independent of social change,[54] motivation for having children has changed little. Here, even more than in industrialized nations, the family has kept on producing abundant offspring, even though only a fraction of these children are now needed.

If excessive population growth is to be prevented, the obvious requirement is somehow to impose restraints on the family. However, because family roles are reinforced by society's system of rewards, punishments, sentiments, and norms, any proposal to demote the family is viewed as a threat by conservatives and liberals alike, and certainly by people with enough social responsibility to work for population control. One is charged with trying to "abolish" the family, but what is required is selective restructuring of the family in relation to the rest of society.

The lines of such restructuring are suggested by two existing limitations on fertility. (i) Nearly all societies succeed in drastically discouraging reproduction among unmarried women. (ii) Advanced societies unintentionally reduce reproduction among married women when conditions worsen in such a way as to penalize childbearing more severely than it was penalized before. In both cases the causes are motivational and economic rather than technological.

It follows that population-control policy can de-emphasize the family in two ways: (i) by keeping present controls over illegitimate childbirth yet making the most of factors that lead people to postpone or avoid marriage, and (ii) by instituting conditions that motivate those who do marry to keep their families small.

Since the female reproductive span is short and generally more fecund in its first than in its second half, postponement of marriage to ages beyond 20 tends biologically to reduce births. Sociologically, it gives women time to get a better education, acquire interests unrelated to the family, and develop a cautious attitude toward pregnancy.[55] Individuals who have not married by the time they are in their late twenties often do not marry at all. For these reasons, for the world as a whole, the average age at marriage for women is negatively associated with the birth rate: a rising age at marriage is a frequent cause of declining fertility during the middle phase of the demographic transition; and, in the late phase, the "baby boom" is usually associated with a return to younger marriages.

Any suggestion that age at marriage be raised as a part of population policy is usually met with the argument that "even if a law were passed, it would not be obeyed." Interestingly, this objection implies that the only way to control the age at marriage is by direct legislation, but other factors govern the actual age. Roman Catholic countries generally follow canon law in stipulating 12 years as the minimum *legal* age at which girls may marry, but the actual average age at marriage in these countries (at least in Europe) is characteristically more like 25 to 28 years. The actual age is determined, not by law, but by social and economic conditions. In agrarian societies, postponement of marriage (when postponement occurs) is apparently caused by difficulties in meeting the economic prerequisites for matrimony, as stipulated by custom and opinion. In industrial societies it is caused by housing shortages, unemployment, the requirement for overseas military service, high costs of education, and inadequacy of consumer services. Since almost no research has been devoted to the subject, it is difficult to assess the relative weight of the factors that govern the age at marriage.

As a means of encouraging the limitation of reproduction within marriage, as well as postponement of marriage, a greater rewarding of nonfamilial than of familial roles would probably help. A simple way of accomplishing this would be to allow economic advantages to accrue to the single as opposed to the married individual, and to the small as opposed to the large family. For instance, the government could pay people to permit themselves to be sterilized;[56] all costs of abortion could be paid by the government; a substantial fee could be charged for a marriage license; a "child-tax"[57] could be levied; and there could be a requirement that illegitimate pregnancies be aborted. Less sensationally, governments could simply reverse some existing policies that encourage childbearing. They could, for example, cease taxing single persons more than married ones; stop giving parents special tax exemptions; abandon income-tax policy that discriminates against couples when the wife works; reduce paid maternity leaves; reduce family allowances;[58] stop awarding public housing on the basis of family size; stop granting fellowships and other educational aids (including special allowances for wives and children) to married students; cease outlawing abortions and sterilizations; and relax rules that allow use of harmless contraceptives only with medical permission. Some of these policy reversals would be beneficial in other than demographic respects and some would be harmful unless special precautions were taken. The aim would be to reduce the number, not the quality, of the next generation.

A closely related method of de-emphasizing the family would be modification of the complementarity of the roles of men and women. Men are now able to participate in the wider world yet enjoy the satisfaction of having several children because the housework and childcare fall mainly on their wives. Women are impelled to seek this role by their idealized view of marriage and motherhood and by either the scarcity of alternative roles or the difficulty of combining them with family roles. To change this situation women could be required to work outside the home, or compelled by circumstances to do so. If, at the same time, women were paid as well as men and given equal educational and occupational opportunities, and if social life were organized around the place of work rather than around the home or neighborhood, many women would develop interests that would compete with family interests. Approximately this policy is now followed in several Communist countries, and even the less developed of these currently have extremely low birth rates.[59]

That inclusion of women in the labor force has a negative effect on reproduction is indicated by regional comparisons.[60,61] But in most countries the wife's employment is subordinate, economically and emotionally, to her family role, and is readily sacrificed for the latter. No society has restructured both the occupational system and the domestic establishment to the point of permanently modifying the old division of labor by sex.

In any deliberate effort to control the birth rate along these lines, a government has two powerful instruments—its command over economic planning and its authority (real or potential) over education. The first determines (as far as policy can) the economic conditions and circumstances affecting the lives of all citizens; the second provides the knowledge and attitudes necessary to implement the plans. The economic system largely determines who shall work, what can be bought, what rearing children will cost, how much individuals can spend. The schools define family roles and develop vocational and recreational interests; they could, if it were desired, redefine the sex roles, develop interests that transcend the home, and transmit realistic (as opposed to moralistic) knowledge concerning marriage, sexual behavior, and population problems. When the problem is viewed in this light, it is clear that the ministries of economics and education, not the ministry of health, should be the source of population policy.

It should now be apparent why, despite strong anxiety over runaway population growth, the actual programs purporting to control it are limited to family planning and are therefore ineffective. (i) The goal of zero, or even slight, population growth is one that nations and groups find difficult to accept. (ii) The measures that would be required to implement such a goal, though not so revolutionary as a Brave New World or a Communist Utopia, nevertheless tend to offend most people reared in existing societies. As a consequence, the goal of so-called population control is implicit and vague; the method is only family planning. This method, far from de-emphasizing the family, is familistic. One of its stated goals is that of helping sterile couples to *have* children. It stresses parental aspirations and responsibilities. It goes along with most aspects of conventional morality, such as condemnation of abortion, disapproval of premarital intercourse, respect for religious teachings and cultural taboos, and obeisance to medical and clerical authority. It deflects hostility by refusing to recommend any change other than the one it stands for: availability of contraceptives.

The things that make family planning acceptable are the very things that make it ineffective for population control. By stressing the right of parents to have the number of children they want, it evades the basic question of population policy, which is how to give societies the number of children they need. By offering only the means for *couples* to control fertility, it neglects the means for societies to do so.

Because of the predominantly pro-family character of existing societies, individual interest ordinarily leads to the production of enough offspring to constitute rapid population growth under conditions of low mortality. Childless or single-child homes are considered indicative of personal failure, whereas having three to five living children gives a family a sense of continuity and substantiality.[62]

Given the existing desire to have moderate-sized rather than small families, the only countries in which fertility has been reduced to match reduction in mortality are advanced ones temporarily experiencing worsened economic conditions. In Sweden, for instance, the net reproduction rate (NRR) has been below replacement for 34 years (1930-63), if the period is taken as a whole, but this is because of the economic depression. The average replacement rate was below unity (NRR = 0.81) for the period 1930-42, but from 1942 through 1963 it was above unity (NRR = 1.08). Hardships that seem particularly conducive to deliberate lowering of the birth rate are (in managed economies) scarcity of housing and other consumer goods despite full employment, and required high participation of women in the labor force, or (in freer economies) a great deal of unemployment and economic insecurity. When conditions are good, any nation tends to have a growing population.

It follows that, in countries where contraception is used, a realistic proposal for a government policy of lowering the birth rate reads like a catalogue of horrors: squeeze consumers through taxation and inflation; make housing very scarce by limiting construction; force wives and mothers to work outside the home to offset the inadequacy of male wages, yet provide few child-care facilities; encourage migration to the city by paying low wages in the country and providing few rural jobs; increase congestion in cities by starving the transit system; increase personal insecurity by encouraging conditions that produce unemployment and by haphazard political arrests. No government will institute such hardships simply for the purpose of controlling population growth. Clearly, therefore, the task of contemporary population policy is to develop attractive substitutes for family interests, so as to avoid having to turn to hardship as a corrective. The specific measures required for developing such substitutes are not easy to determine in the absence of research on the question.

In short, the world's population problem cannot be solved by pretense and wishful thinking. The unthinking identification of family planning with population control is an ostrich-like approach in that it permits people to hide from themselves the enormity and unconventionality of the task. There is no reason to abandon family-planning programs; contraception is a valuable technological instrument. But such programs must be supplemented with equal or greater investments in research and experimentation to determine the required socioeconomic measures.

Sir Julian Huxley spoke somewhat optimistically of "family planning" as an effective means of controlling population growth. In the following article, Kingsley Davis (1908-   ) examines with a skeptical eye the philosophy behind family planning. He sees the possibility of disastrous demographic results from programs which let individual couples decide how many children they will have. Most people, says Davis, simply want too many children, and their progeny, planned for or not, contribute to the population explosion. In addition to his skepticism about planning for children, Davis questions the related emphasis upon the family and calls for encouragement of postponement of marriage, for a limit to the number of births allowed per marriage, for government-subsidized sterilization and abortion programs, for taxes on children, etc. Such socioeconomic approaches, according to Davis, must supplement our exclusive reliance on family planning as a method of population control.

Kingsley Davis is a professor of sociology and director of International Population and Urban Research at the University of California at Berkeley.

THE TRAGEDY OF THE COMMONS

by Garrett Hardin

At the end of a thoughtful article on the future of nuclear war, Wiesner and York concluded that: "Both sides in the arms race are . . . confronted by the dilemma of steadily increasing military power and steadily decreasing national security. *It is our considered professional judgment that this dilemma has no technical solution.* If the great powers continue to look for solutions in the area of science and technology only, the result will be to worsen the situation."[1]

I would like to focus your attention not on the subject of the article (national security in a nuclear world) but on the kind of conclusion they reached, namely that there is no technical solution to the problem. An implicit and almost universal assumption of discussions published in professional and semipopular scientific journals is that the problem under discussion has a technical solution. A technical solution may be defined as one that requires a change only in the techniques of the natural sciences, demanding little or nothing in the way of change in human values or ideas of morality.

"The Tragedy of the Commons" From *Science*, December 13, 1968, Vol. 162, pp. 1243-1248.

In our day (though not in earlier times) technical solutions are always welcome. Because of previous failures in prophecy, it takes courage to assert that a desired technical solution is not possible. Wiesner and York exhibited this courage; publishing in a science journal, they insisted that the solution to the problem was not to be found in the natural sciences. They cautiously qualified their statement with the phrase, "It is our considered professional judgment. . . ." Whether they were right or not is not the concern of the present article. Rather, the concern here is with the important concept of a class of human problems which can be called "no technical solution problems," and, more specifically, with the identification and discussion of one of these.

It is easy to show that the class is not a null class. Recall the game of tick-tack-toe. Consider the problem, "How can I win the game of tick-tack-toe?" It is well known that I cannot, if I assume (in keeping with the conventions of game theory) that my opponent understands the game perfectly. Put another way, there is no "technical solution" to the problem. I can win only by giving a radical meaning to the word "win." I can hit my opponent over the head; or I can drug him; or I can falsify the records. Every way in which I "win" involves, in some sense, an abandonment of the game, as we intuitively understand it. (I can also, of course, openly abandon the game—refuse to play it. This is what most adults do.)

The class of "No technical solution problems" has members. My thesis is that the "population problem," as conventionally conceived, is a member of this class. How it is conventionally conceived needs some comment. It is fair to say that most people who anguish over the population problem are trying to find a way to avoid the evils of overpopulation without relinquishing any of the privileges they now enjoy. They think that farming the seas or developing new strains of wheat will solve the problem—technologically. I try to show here that the solution they seek cannot be found. The population problem cannot be solved in a technical way, any more than can the problem of winning the game of tick-tack-toe.

Population, as Malthus said, naturally tends to grow "geometrically," or, as we would now say, exponentially. In a finite world this means that the per capita share of the world's goods must steadily decrease. Is ours a finite world?

A fair defense can be put forward for the view that the world is infinite; or that we do not know that it is not. But, in terms of the practical problems that we must face in the next few generations with the foreseeable technology, it is clear that we will greatly increase human misery if we do not, during the immediate future, assume that the world available to the terrestrial human population is finite. "Space" is no escape.[2]

A finite world can support only a finite population; therefore, population growth must eventually equal zero. (The case of perpetual wide fluctuations above and below zero is a trivial variant that need not be discussed.) When this condition is met, what will be the situation of mankind? Specifically, can Bentham's goal of "the greatest good for the greatest number" be realized?

No—for two reasons, each sufficient by itself. The first is a theoretical one. It is not mathematically possible to maximize for two (or more) variables at the same time. This was clearly stated by von Neumann and Morgenstern,[3] but the principle is implicit in the theory of partial differential equations, dating back at least to D'Alembert (1717-1783).

The second reason springs directly from biological facts. To live, any organism must have a source of energy (for example, food). This energy is utilized for two purposes: mere maintenance and work. For man, maintenance of life requires about 1600 kilo-calories a day ("maintenance calories"). Anything that he does over and above merely staying alive will be defined as work, and is supported by "work calories" which he takes in. Work calories are used not only for what we call work in common speech; they are also required for all forms of enjoyment, from swimming and automobile racing to playing music and writing poetry. If our goal is to maximize population it is obvious what we must do: We must make the work calories per person approach as close to zero as possible. No gourmet meals, no vacations, no sports, no music, no literature, no art. . . . I think that everyone will grant, without argument or proof, that maximizing population does not maximize goods. Bentham's goal is impossible.

In reaching this conclusion I have made the usual assumption that it is the acquisition of energy that is the problem. The appearance of atomic energy has led some to question this assumption. However, given an infinite source of energy, population growth still produces an inescapable problem. The problem of the acquisition of energy is replaced by the problem of its dissipation, as J. H. Fremlin has so wittily shown.[4] The arithmetic signs in the analysis are, as it were, reversed; but Bentham's goal is still unobtainable.

The optimum population is, then, less than the maximum. The difficulty of defining the optimum is enormous; so far as I know, no one has seriously tackled this problem. Reaching an acceptable and stable solution will surely require more than one generation of hard analytical work—and much persuasion.

We want the maximum good per person; but what is good? To one person it is wilderness, to another it is ski lodges for thousands. To one it is estuaries to nourish ducks for hunters to shoot; to another it is factory land. Comparing one good with another is, we usually say, impossible because goods are incommensurable. Incommensurables cannot be compared.

Theoretically this may be true; but in real life incommensurables *are* commensurable. Only a criterion of judgment and a system of weighting are needed. In nature the criterion is survival. Is it better for a species to be small and hideable, or large and powerful? Natural selection commensurates the incommensurables. The compromise achieved depends on a natural weighting of the values of the variables.

Man must imitate this process. There is no doubt that in fact he already does, but unconsciously. It is when the hidden decisions are made explicit that the arguments begin. The problem for the years ahead is to work out an acceptable theory of weighting. Synergistic effects, nonlinear variation, and difficulties in discounting the future make the intellectual problem difficult, but not (in principle) insoluble.

Has any cultural group solved this practical problem at the present time, even on an intuitive level? One simple fact proves that none has: there is no prosperous population in the world today that has, and has had for some time, a growth rate of zero. Any people that has intuitively identified its optimum point will soon reach it, after which its growth rate becomes and remains zero.

Of course, a positive growth rate might be taken as evidence that a population is below its optimum. However, by any reasonable standards, the most rapidly growing populations on earth today are (in general) the most miserable. This association (which need not be invariable) casts doubt on the optimistic assumption that the positive growth rate of a population is evidence that it has yet to reach its optimum.

We can make little progress in working toward optimum population size until we explicitly exorcize the spirit of Adam Smith in the field of practical demography. In economic affairs, *The Wealth of Nations* (1776) popularized the "invisible hand," the idea that an individual who "intends only his own gain," is, as it were, "led by an invisible hand to promote . . . the public interest."[5] Adam Smith did not assert that this was invariably true, and perhaps neither did any of his followers. But he contributed to a dominant tendency of thought that has ever since interfered with positive action based on rational analysis, namely, the tendency to assume that decisions reached individually will, in fact, be the best decisions for an entire society. If this assumption is correct it justifies the continuance of our present policy of laissez-faire in reproduction. If it is correct we can assume that men will control their individual fecundity so as to produce the optimum population. If the assumption is not correct, we need to reexamine our individual freedoms to see which ones are defensible.

The rebuttal to the invisible hand in population control is to be found in a scenario first sketched in a little-known pamphlet in 1833 by a mathematical amateur named William Forster Lloyd (1794-1852).[6] We may well call it "the tragedy of the commons," using the word "tragedy" as the philosopher Whitehead used it: "The essence of dramatic tragedy is not unhappiness. It resides in the solemnity of the remorseless working of things."[7] He then goes on to say, "This inevitableness of destiny can only be illustrated in terms of human life by incidents which in fact involve unhappiness. For it is only by them that the futility of escape can be made evident in the drama."

The tragedy of the commons develops in this way. Picture a pasture open to all. It is to be expected that each herdsman will try to keep as many cattle as possible on the commons. Such an arrangement may work reasonably satisfactorily for centuries because tribal wars, poaching, and disease keep the numbers of both man and beast well below the carrying capacity of the land. Finally, however, comes the day of reckoning, that is, the day when the long-desired goal of social stability becomes a reality. At this point, the inherent logic of the commons remorselessly generates tragedy.

As a rational being, each herdsman seeks to maximize his gain. Explicitly or implicitly, more or less consciously, he asks, "What is the utility *to me* of adding one more animal to my herd?" This utility has one negative and one positive component.

1) The positive component is a function of the increment of one animal. Since the herdsman receives all the proceeds from the sale of the additional animal, the positive utility is nearly +1.

2) The negative component is a function of the additional overgrazing created by one more animal. Since, however, the effects of overgrazing are shared by all the herdsmen, the negative utility for any particular decision-making herdsman is only a fraction of -1.

Adding together the component partial utilities, the rational herdsman concludes that the only sensible course for him to pursue is to add another animal to his herd. And another; and another. . . . But this is the conclusion reached by each and every rational herdsman sharing a commons. Therein is the tragedy. Each man is locked into a system that compels him to increase his herd without limit—in a world that is limited. Ruin is the destination toward which all men rush, each pursuing his own best interest in a society that believes in the freedom of the commons. Freedom in a commons brings ruin to all.

Some would say that this is a platitude. Would that it were! In a sense, it was learned thousands of years ago, but natural selection favors the forces of psychological denial.[8] The individual benefits as an individual from his ability to deny the truth even though society as a whole, of which he is a part, suffers. Education can counteract the natural tendency to do the wrong thing, but the inexorable succession of generations requires that the basis for this knowledge be constantly refreshed.

A simple incident that occurred a few years ago in Leominster, Massachusetts, shows how perishable the knowledge is. During the Christmas shopping season the parking meters downtown were covered with plastic bags that bore tags reading: "Do not open until after Christmas. Free parking courtesy of the mayor and city council." In other words, facing the prospect of an increased demand for already scarce space, the city fathers reinstituted the system of the commons. (Cynically, we suspect that they gained more votes than they lost by this retrogressive act.)

In an approximate way, the logic of the commons has been understood for a long time, perhaps since the discovery of agriculture or the invention of private property in real estate. But it is understood mostly only in special cases which are not sufficiently generalized. Even at this late date, cattlemen leasing national land on the western ranges demonstrate no more than an ambivalent understanding, in constantly pressuring federal authorities to increase the head count to the point where overgrazing produces erosion and weed-dominance. Likewise, the oceans of the world continue to suffer from the survival of the philosophy of the commons. Maritime nations still respond automatically to the shibboleth of the "freedom of the seas." Professing to believe in the "inexhaustible resources of the oceans," they bring species after species of fish and whales closer to extinction.[9]

The National Parks present another instance of the working out of the tragedy of the commons. At present, they are open to all, without limit. The parks themselves are limited in extent—there is only one Yosemite Valley—whereas population seems to grow without limit. The values that visitors seek in the parks are steadily eroded. Plainly, we must soon cease to treat the parks as commons or they will be of no value to anyone.

What shall we do? We have several options. We might sell them off as private property. We might keep them as public property, but allocate the right to enter them. The allocation might be on the basis of wealth, by the use of an auction system. It might be on the basis of merit, as defined by some agreed-upon standards. It might be by lottery. Or it might be on a first-come, first-served basis, administered to long queues. These, I think, are all the reasonable possibilities. They are all objectionable. But we must choose—or acquiesce in the destruction of the commons that we call our National Parks.

In a reverse way, the tragedy of the commons reappears in problems of pollution. Here it is not a question of taking something out of the commons, but of putting something in—sewage, or chemical, radioactive, and heat wastes into water; noxious and dangerous fumes into the air; and distracting and unpleasant advertising signs into the line of sight. The calculations of utility are much the same as before. The rational man finds that his share of the cost of the wastes he discharges into the commons is less than the cost of purifying his wastes before releasing them. Since this is true for everyone, we are locked into a system of "fouling our own nest," so long as we behave only as independent, rational, free-enterprisers.

The tragedy of the commons as a food basket is averted by private property, or something formally like it. But the air and waters surrounding us cannot readily be fenced, and so the tragedy of the commons as a cesspool must be prevented by different means, by coercive laws or taxing devices that make it cheaper for the polluter to treat his pollutants than to discharge them untreated. We have not progressed as far with the solution of this problem as we have with the first. Indeed, our particular concept of private property, which deters us from exhausting the positive resources of the earth, favors pollution. The owner of a factory on the bank of a stream—whose property extends to the middle of the stream—often has difficulty seeing why it is not his natural right to muddy the waters flowing past his door. The law, always behind the times, requires elaborate stitching and fitting to adapt it to this newly perceived aspect of the commons.

The pollution problem is a consequence of population. It did not much matter how a lonely American frontiersman disposed of his waste. "Flowing water purifies itself every 10 miles," my grandfather used to say, and the myth was near enough to the truth when he was a boy, for there were not too many people. But as population became denser, the natural chemical and biological recycling processes became overloaded, calling for a redefinition of property rights.

Analysis of the pollution problem as a function of population density uncovers a not generally recognized principle of morality, namely: *the morality of an act is a function of the state of the system at the time it is performed.*[10] Using the commons as a cesspool does not harm the general public under frontier conditions, because there is no public; the same behavior in a metropolis is unbearable. A hundred and fifty years ago a plainsman could kill an American bison, cut out only the tongue for his dinner, and discard the rest of the animal. He was not in any important sense being wasteful. Today, with only a few thousand bison left, we would be appalled at such behavior.

In passing, it is worth noting that the morality of an act cannot be determined from a photograph. One does not know whether a man killing an elephant or setting fire to the grassland is harming others until one knows the total system in which his act appears. "One picture is worth a thousand words," said an ancient Chinese; but it may take 10,000 words to validate it. It is as tempting to ecologists as it is to reformers in general to try to persuade others by way of the photographic shortcut. But the essence of an argument cannot be photographed: it must be presented rationally—in words.

That morality is system-sensitive escaped the attention of most codifiers of ethics in the past. "Thou shalt not . . ." is the form of traditional ethical directives which make no allowance for particular circumstances. The laws of our society follow the pattern of ancient ethics, and therefore are poorly suited to governing a complex, crowded, changeable world. Our epicyclic solution is to augment statutory law with administrative law. Since it is practically impossible to spell out all the conditions under which it is safe to burn trash in the back yard or to run an automobile without smog-control, by law we delegate the details to bureaus. The result is administrative law, which is rightly feared for an ancient reason—*Quis custodiet ipsos custodes?*—"Who shall watch the watchers themselves?" John Adams said that we must have "a government of laws and not men." Bureau administrators, trying to evaluate the morality of acts in the total system, are singularly liable to corruption, producing a government by men, not laws.

Prohibition is easy to legislate (though not necessarily to enforce); but how do we legislate temperance? Experience indicates that it can be accomplished best through the mediation of administrative law. We limit possibilities unnecessarily if we suppose that the sentiment of *Quis custodiet* denies us the use of administrative law. We should rather retain the phrase as a perpetual reminder of fearful dangers we cannot avoid. The great challenge facing us now is to invent the corrective feedbacks that are needed to keep custodians honest. We must find ways to legitimate the needed authority of both the custodians and the corrective feedbacks.

The tragedy of the commons is involved in population problems in another way. In a world governed solely by the principle of "dog eat dog"—if indeed there was such a world—how many children a family had would not be a matter of public concern. Parents who bred too exuberantly would leave fewer descendants, not more, because they would be unable to care adequately for their children. David Lack and others have found that such a negative feedback demonstrably controls the fecundity of birds.[11] But men are not birds, and have not acted like them for millenniums, at least.

If each human family were dependent only on its own resources; *if* the children of improvident parents starved to death; *if,* thus, overbreeding brought its own "punishment" to the germ line—*then* there would be no public interest in controlling the breeding of families. But our society is deeply committed to the welfare state,[12] and hence is confronted with another aspect of the tragedy of the commons.

In a welfare state, how shall we deal with the family, the religion, the race, or the class (or indeed any distinguishable and cohesive group) that adopts overbreeding as a policy to secure its own aggrandizement?[13] To couple the concept of freedom to breed with the belief that everyone born has an equal right to the commons is to lock the world into a tragic course of action.

Unfortunately this is just the course of action that is being pursued by the United Nations. In late 1967, some 30 nations agreed to the following:

"The Universal Declaration of Human Rights describes the family as the natural and fundamental unit of society. It follows that any choice and decision with regard to the size of the family must irrevocably rest with the family itself, and cannot be made by anyone else."[14]

It is painful to have to deny categorically the validity of this right; denying it, one feels as uncomfortable as a resident of Salem, Massachusetts, who denied the reality of witches in the 17th century. At the present time, in liberal quarters, something like a taboo acts to inhibit criticism of the United Nations. There is a feeling that the United Nations is "our last and best hope," that we shouldn't find fault with it; we shouldn't play into the hands of the archconservatives. However, let us not forget what Robert Louis Stevenson said: "The truth that is suppressed by friends is the readiest weapon of the enemy." If we love the truth we must openly deny the validity of the Universal Declaration of Human Rights, even though it is promoted by the United Nations. We should also join with Kingsley Davis[15] in attempting to get Planned Parenthood-World Population to see the error of its ways in embracing the same tragic ideal.

It is a mistake to think that we can control the breeding of mankind in the long run by an appeal to conscience. Charles Galton Darwin made this point when he spoke on the centennial of the publication of his grandfather's great book. The argument is straightforward and Darwinian.

People vary. Confronted with appeals to limit breeding, some people will undoubtedly respond to the plea more than others. Those who have more children will produce a larger fraction of the next generation than those with more susceptible consciences. The difference will be accentuated, generation by generation.

In C. G. Darwin's words: "It may well be that it would take hundreds of generations for the progenitive instinct to develop in this way, but if it should do so, nature would have taken her revenge, and the variety *Homo contracipiens* would become extinct and would be replaced by the variety *Homo progenitivus*."[16]

The argument assumes that conscience or the desire for children (no matter which) is hereditary—but hereditary only in the most general formal sense. The result will be the same whether the attitude is transmitted through germ cells, or exosomatically, to use A. J. Lotka's term. (If one denies the latter possibility as well as the former, then what's the point of education?) The argument has here been stated in the context of the population problem, but it applies equally well to any instance in which society appeals to an individual exploiting a commons to restrain himself for the general good—by means of his conscience. To make such an appeal is to set up a selective system that works toward the elimination of conscience from the race.

The long-term disadvantage of an appeal to conscience should be enough to condemn it; but has serious short-term disadvantages as well. If we ask a man who is exploiting a commons to desist "in the name of conscience," what are we saying to him? What does he hear?—not only at the moment but also in the wee small hours of the night when, half asleep, he remembers not merely the words we used but also the nonverbal communication cues we gave him unawares? Sooner or later, consciously or subconsciously, he senses that he has received two communications, and that they are contradictory: (i) (intended communication) "If you don't do as we ask, we will openly condemn you for not acting like a responsible citizen"; (ii) (the unintended communication) "If you *do* behave as we ask, we will secretly condemn you for a simpleton who can be shamed into standing aside while the rest of us exploit the commons."

Everyman then is caught in what Bateson has called a "double bind." Bateson and his co-workers have made a plausible case for viewing the double bind as an important causative factor in the genesis of schizophrenia.[17] The double bind may not always be so damaging, but it always endangers the mental health of anyone to whom it is applied. "A bad conscience," said Nietzsche, "is a kind of illness."

To conjure up a conscience in others is tempting to anyone who wishes to extend his control beyond the legal limits. Leaders at the highest level succumb to this temptation. Has any President during the past generation failed to call on labor unions to moderate voluntarily their demands for higher wages, or to steel companies to honor voluntary guidelines on prices? I can recall none. The rhetoric used on such occasions is designed to produce feelings of guilt in noncooperators.

For centuries it was assumed without proof that guilt was a valuable, perhaps even an indispensable, ingredient of the civilized life. Now, in this post-Freudian world, we doubt it.

Paul Goodman speaks from the modern point of view when he says: "No good has ever come from feeling guilty, neither intelligence, policy, nor compassion. The guilty do not pay attention to the object but only to themselves, and not even to their own interests, which might make sense, but to their anxieties."[18]

One does not have to be a professional psychiatrist to see the consequences of anxiety. We in the Western world are just emerging from a dreadful two-centuries-long Dark Ages of Eros that was sustained partly by prohibition laws, but perhaps more effectively by the anxiety-generating mechanisms of education. Alex Comfort has told the story well in *The Anxiety Makers;*[19] it is not a pretty one.

Since proof is difficult, we may even concede that the results of anxiety may sometimes, from certain points of view, be desirable. The larger question we should ask is whether, as a matter of policy, we should ever encourage the use of a technique the tendency (if not the intention) of which is psychologically pathogenic. We hear much talk these days of responsible parenthood; the coupled words are incorporated into the titles of some organizations devoted to birth control. Some people have proposed massive propaganda campaigns to instill responsibility into the nation's (or the world's) breeders. But what is the meaning of the word responsibility in this context; Is it not merely a synonym for the word conscience? When we use the word responsibility in the absence of substantial sanctions are we not trying to browbeat a free man in a commons into acting against his own interest? Responsibility is a verbal counterfeit for a substantial *quid pro quo*. It is an attempt to get something for nothing.

If the word responsibility is to be used at all, I suggest that it be in the sense Charles Frankel uses it.[20] "Responsibility," says this philosopher, "is the product of definite social arrangements." Notice that Frankel calls for social arrangements—not propaganda.

The social arrangements that produce responsibility are arrangements that create coercion, of some sort. Consider bank-robbing. The man who takes money from a bank acts as if the bank were a commons. How do we prevent such action? Certainly not by trying to control his behavior solely by a verbal appeal to his sense of responbility. Rather than rely on propaganda we follow Frankel's lead and insist that a bank is not a commons; we seek the definite social arrangements that will keep it from becoming a commons. That we thereby infringe on the freedom of would-be robbers we neither deny nor regret.

The morality of bank-robbing is particularly easy to understand because we accept complete prohibition of this activity. We are willing to say "Thou shalt not rob banks," without providing for exceptions. But temperance also can be created by coercion. Taxing is a good coercive device. To keep downtown shoppers temperate in their use of parking space we introduce parking meters for short periods, and traffic fines for longer ones. We need not actually forbid a citizen to park as long as he wants to; we need merely make it increasingly expensive for him to do so. Not prohibition, but carefully biased options are what we offer him. A Madison Avenue man might call this persuasion; I prefer the greater candor of the word coercion.

Coercion is a dirty word to most liberals now, but it need not forever be so. As with the four-letter words, its dirtiness can be cleansed away by exposure to the light, by saying it over and over without apology or embarrassment. To many the word coercion implies arbitrary decisions of distant and irresponsible bureaucrats; but this is not a necessary part of its meaning. The only kind of coercion I recommend is mutual coercion, mutually agreed upon by the majority of the people affected.

To say that we mutually agree to coercion is not to say that we are required to enjoy it, or even to pretend we enjoy it. Who enjoys taxes? We all grumble about them. But we accept compulsory taxes because we recognize that voluntary taxes would favor the conscienceless. We institute and (grumblingly) support taxes and other coercive devices to escape the horror of the commons.

An alternative to the commons need not be perfectly just to be preferable. With real estate and other material goods, the alternative we have chosen is the institution of private property coupled with legal inheritance. Is this system perfectly just? As a genetically trained biologist I deny that it is. It seems to me that, if there are to be differences in individual inheritance, legal possession should be perfectly correlated with biological inheritance —that those who are biologically more fit to be the custodians of property and power should legally inherit more. But genetic recombination continually makes a mockery of the doctrine of "like father, like son" implicit in our laws of legal inheritance. An idiot can inherit millions, and a trust fund can keep his estate intact. We must admit that our legal system of private property plus inheritance is unjust—but we put up with it because we are not convinced, at the moment, that anyone has invented a better system. The alternative of the commons is too horrifying to contemplate. Injustice is preferable to total ruin.

It is one of the peculiarities of the warfare between reform and the status quo that it is thoughtlessly governed by a double standard. Whenever a reform measure is proposed it is often defeated when its opponents triumphantly discover a flaw in it. As Kingsley Davis has pointed out,[21] worshippers of the status quo sometimes imply that no reform is possible without unanimous agreement, an implication contrary to historical fact. As nearly as I can make out, automatic rejection of proposed reforms is based on one of two unconscious assumptions: (i) that the status quo is perfect; or (ii) that the choice we face is between reform and no action; if the proposed reform is imperfect, we presumably should take no action at all, while we wait for a perfect proposal.

But we can never do nothing. That which we have done for thousands of years is also action. It also produces evils. Once we are aware that the status quo is action, we can then compare its discoverable advantages and disadvantages with the predicted advantages and disadvantages of the proposed reform, discounting as best we can for our lack of experience. On the basis of such a comparison, we can make a rational decision which will not involve the unworkable assumption that only perfect systems are tolerable.

Perhaps the simplest summary of this analysis of man's population problems is this: the commons, if justifiable at all, is justifiable only under conditions of low-population density. As the human population has increased, the commons has had to be abandoned in one aspect after another.

First we abandoned the commons in food gathering, enclosing farm land and restricting pastures and hunting and fishing areas. These restrictions are still not complete throughout the world.

Somewhat later we saw that the commons as a place for waste disposal would also have to be abandoned. Restrictions on the disposal of domestic sewage are widely accepted in the Western world; we are still struggling to close the commons to pollution by automobiles, factories, insecticide sprayers, fertilizing operations, and atomic energy installations.

In a still more embryonic state is our recognition of the evils of the commons in matters of pleasure. There is almost no restriction on the propagation of sound waves in the public medium. The shopping public is assaulted with mindless music, without its consent. Our government is paying out billions of dollars to create supersonic transport which will disturb 50,000 people for every one person who is whisked from coast to coast 3 hours faster. Advertisers muddy the airwaves of radio and television and pollute the view of travelers. We are a long way from outlawing the commons in matters of pleasure. Is this because our Puritan inheritance makes us view pleasure as something of a sin, and pain (that is, the pollution of advertising) as the sign of virtue?

Every new enclosure of the commons involves the infringement of somebody's personal liberty. Infringements made in the distant past are accepted because no contemporary complains of a loss. It is the newly proposed infringements that we vigorously oppose; cries of "rights" and "freedom" fill the air. But what does "freedom" mean? When men mutually agreed to pass laws against robbing, mankind became more free, not less so. Individuals locked into the logic of the commons are free only to bring on universal ruin; once they see the necessity of mutual coercion, they become free to pursue other goals. I believe it was Hegel who said "Freedom is the recognition of necessity."

The most important aspect of necessity that we must now recognize, is the necessity of abandoning the commons in breeding. No technical solution can rescue us from the misery of overpopulation. Freedom to breed will bring ruin to all. At the moment, to avoid hard decisions many of us are tempted to propagandize for conscience and responsible parenthood. The temptation must be resisted, because an appeal to independently acting consciences selects for the disappearance of all conscience in the long run, and an increase in anxiety in the short.

The only way we can preserve and nurture other and more precious freedoms is by relinquishing the freedom to breed, and that very soon. "Freedom is the recognition of necessity"—and it is the role of education to reveal to all the necessity of abandoning the freedom to breed. Only so, can we put an end to this aspect of the tragedy of the commons.

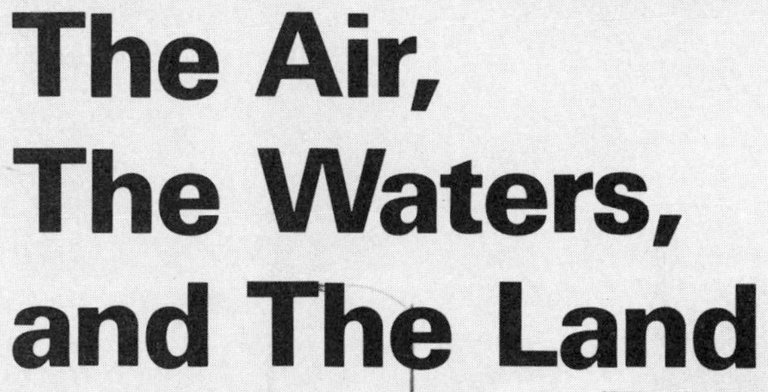

The Air, The Waters, and The Land

This is the first in a group of essays which specifically detail the destruction of our environment through pollution and disregard for ecological principles. It is fitting to begin this section with a description of a killer fog since air pollution is not merely a future threat but is a clear and immediate danger; in fact, the first killer fog occurred in an industrial area in Belgium forty years ago. Such deadly atmospheric conditions can now be expected with increasing frequency as Paul Ehrlich states in "Eco—Catastrophe!" (page 3). Even today school children in Los Angeles are forbidden to exercise or play, i.e., *breathe deeply, on days of heavy smog. "The Fog," winner of the 1950 Lasker Medical Journalism Award, reports the story of America's first killer fog, a six-day catastrophe which took the lives of twenty people in Donora, Pennsylvania, in October 1948.*

Berton Roueché (1911-) is a staff writer for The New Yorker, *in which this article originally appeared. A series of similar narratives of medical detection by Roueché comprise his well-known book,* Eleven Blue Men.

THE FOG

by Berton Roueché

The Monongahela River rises in the middle Alleghenies and seeps for a hundred and twenty-eight miles through the iron and bituminous-coal fields of northeastern West Virginia and southwestern Pennsylvania to Pittsburgh. There, joining the Allegheny River, it becomes the wild Ohio. It is the only river of any consequence in the United States that flows due north, and it is also the shortest. Its course is cramped and crooked, and flanked by bluffs and precipitous hills. Within living memory, its waters were quick and green, but they are murky now with pollution, and a series of locks and dams steady its once tumultuous descent, rendering it navigable from source to mouth. Traffic on the Monongahela is heavy. Its shipping, which consists almost wholly of coal barges pushed by wheezy, coal-burning stern-wheelers, exceeds in tonnage that of the Panama Canal. The river is densely industrialized. There are trucking highways along its narrow banks and interurban lines and branches of the Pennsylvania Railroad and the New York Central and smelters and steel plants and chemical works and glass factories and foundries and coke plants and machine shops and zinc mills, and its hills and bluffs are scaled by numerous blackened mill towns. The blackest of them is the borough of Donora, in Washington County, Pennsylvania.

Donora is twenty-eight miles south of Pittsburgh and covers the tip of a lumpy point formed by the most convulsive of the Monongahela's many horseshoe bends. Though accessible by road, rail, and river, it is an extraordinarily secluded place. The river and the bluffs that lift abruptly from the water's edge to a height of four hundred and fifty feet enclose it on the north and east and south, and just above it to the west is a range of rolling but even higher hills. On its outskirts are acres of sidings and rusting gondolas, abandoned mines, smoldering slag piles, and gulches filled with rubbish. Its limits are marked by sooty signs that read, "Donora. Next to Yours the Best Town in the U.S.A." It is a harsh, gritty town, founded in 1901 and old for its age, with a gaudy main street and a thousand identical gaunt gray houses. Some of its streets are paved with concrete and some are cobbled, but many are of dirt and crushed coal. At least half of them are as steep as roofs, and several have steps instead of sidewalks. It is treeless and all but grassless, and much of it is slowly sliding downhill. After a rain, it is a smear of mud. Its vacant lots and many of its yards are mortally gullied, and one of its three cemeteries is an eroded ruin of gravelly clay and toppled tombstones. Its population is 12,300. Two-thirds of its men, and a substantial number of its women, work in its mills. There are three of them —a steel plant, a wire plant, and a zinc-and-sulphuric-acid plant—all of which are operated by the American Steel & Wire Co., a subsidiary of the United States Steel Corporation, and they line its river front for three miles. They are huge mills. Some of the buildings are two blocks long, many are five or six stories high, and all of them bristle with hundred-foot stacks perpetually plumed with black or red or sulphurous yellow smoke.

Donora is abnormally smoky. Its mills are no bigger or smokier than many, but their smoke, and the smoke from the passing boats and trains, tends to linger there. Because of the crowding bluffs and sheltering hills, there is seldom a wind, and only occasionally a breeze, to dispel it. On still days, unless the skies are high and buoyantly clear, the lower streets are always dim and there is frequently a haze on the heights. Autumn is the smokiest season. The weather is close and dull then, and there are persistent fogs as well. The densest ones generally come in October. They are greasy, gagging fogs, often intact even at high noon, and they sometimes last for two or three days. A few have lasted as long as four. One, toward the end of October, 1948, hung on for six. Unlike its predecessors, it turned out to be of considerably more than local interest. It was the second smoke-contaminated fog in history ever to reach a toxic density. The first such fog occurred in Belgium, in an industrialized stretch of the Meuse Valley, in 1930. During it several hundred people were prostrated, sixty of them fatally. The Donora fog struck down nearly six thousand. Twenty of them—five women and fifteen men—died. Nobody knows exactly what killed them, or why the others survived. At the time, not many of the stricken expected to.

The fog closed over Donora on the morning of Tuesday, October 26th. The weather was raw, cloudy, and dead calm, and it stayed that way as the fog piled up all that day and the next. By Thursday, it had stiffened adhesively into a motionless clot of smoke. That afternoon, it was just possible to see across the street, and, except for the stacks, the mills had vanished. The air began to have a sickening smell, almost a taste. It was the bittersweet reek of sulphur dioxide. Everyone who was out that day remarked on it, but no one was much concerned. The smell of sulphur dioxide, a scratchy gas given off by burning coal and melting ore, is a normal concomitant of any durable fog in Donora. This time, it merely seemed more penetrating than usual.

At about eight-thirty on Friday morning, one of Donora's eight physicians, Dr. Ralph W. Koehler, a tense, stocky man of forty-eight, stepped to his bathroom window for a look at the weather. It was, at best, unchanged. He could see nothing but a watery waste of rooftops islanded in fog. As he was turning away, a shimmer of movement in the distance caught his eye. It was a freight train creeping along the riverbank just south of town, and the sight of it shook him. He had never seen anything quite like it before. "It was the smoke," he says. "They were firing up for the grade and the smoke was belching out, but it didn't rise. I mean it didn't go up at all. It just spilled out over the lip of the stack like a black liquid, like ink or oil, and rolled down to the ground and lay there. My God, it just lay there! I thought, Well, God damn—and they talk about needing smoke control up in Pittsburgh! I've got a heart condition, and I was so disgusted my heart began to act up a little. I had to sit down on the edge of the tub and rest a minute.

Dr. Koehler and an associate, Dr. Edward Roth, who is big, heavyset, and in his middle forties, share an office on the second floor of a brownstone building one block up from the mills, on McKean Avenue, the town's main street. They have one employee, a young woman named Helen Stack, in whom are combined an attractive receptionist, an efficient secretary, and a capable nurse. Miss Stack was the first to reach the office that morning.

Like Dr. Koehler and many other Donorans, she was in uncertain spirits. The fog was beginning to get on her nerves, and she had awakened with a sore throat and a cough and supposed that she was coming down with a cold. The appearance of the office deepened her depression. Everything in it was smeared with a kind of dust. "It wasn't just ordinary soot and grit," she says. "There was something white and scummy mixed up in it. It was just wet ash from the mills, but I didn't know that then. I almost hated to touch it, it was so nasty-looking. But it had to be cleaned up, so I got out a cloth and went to work." When Miss Stack had finished, she lighted a cigarette and sat down at her desk to go through the mail. It struck her that the cigarette had a very peculiar taste. She held it up and sniffed at the smoke. Then she raised it to her lips, took another puff, and doubled up in a paroxysm of coughing. For an instant, she thought she was going to be sick. "I'll never forget that taste," she says. "Oh, it was awful! It was sweet and horrible, like something rotten. It tasted the way the fog smelled, only ten times worse. I got rid of the cigarette as fast as I could and drank a glass of water, and then I felt better. What puzzled me was I'd smoked a cigarette at home after breakfast and it had tasted all right. I didn't know what to think, except that maybe it was because the fog wasn't quite as bad up the hill as here downstreet. I guess I thought my cold was probably partly to blame. I wasn't really uneasy. The big Halloween parade the Chamber of Commerce puts on every year was to be held that night, and I could hear the workmen down in the street putting up the decorations. I knew the committee wouldn't be going ahead with the parade if they thought anything was wrong. So I went on with my work, and pretty soon the Doctors came in from their early calls and it was just like any other morning."

The office hours of Dr. Koehler and Dr. Roth are the same, from one to three in the afternoon and from seven to nine at night. Whenever possible in the afternoon, Dr. Koehler leaves promptly at three. Because of his unsteady heart, he finds it desirable to rest for a time before dinner. That Friday afternoon, he was just getting into his coat when Miss Stack announced a patient. "He was wheezing and gasping for air," Dr. Koehler says, "but there wasn't anything very surprising about that. He was one of our regular asthmatics, and the fog gets them every time. The only surprising thing was that he hadn't come in sooner. The fact is, none of our asthmatics had been in all week. Well, I did what I could for him. I gave him a shot of adrenalin or aminophyllin—some anti-spasmodic—to dilate the bronchia, so he could breathe more easily, and sent him home. I followed him out. I didn't feel so good myself."

Half an hour after Dr. Koehler left, another gasping asthmatic, an elderly steelworker, tottered into the office. "He was pretty wobbly," Miss Stack says. "Dr. Roth was still in his office, and saw him right away. I guess he wasn't much better when he came out, because I remember thinking, Poor fellow. There's nothing sadder than an asthmatic when the fog is bad. Well, he had hardly gone out the door when I heard a terrible commotion. I thought, Oh, my gosh, he's fallen down the stairs! Then there was an awful yell. I jumped up and dashed out into the hall. There was a man I'd never seen before sort of draped over the bannister. He was kicking at the wall and pulling at the banister and moaning and choking and yelling at the top of his voice, 'Help! Help me! I'm dying!' I just stood there. I was petrified. Then Dr. Brown, across the hall, came running out, and he and somebody else helped the man on up the stairs and into his office. Just then, my phone began to ring. I almost bumped into Dr. Roth. He was coming out to see what was going on. When I picked up the phone, it was just like hearing that man in the hall again. It was somebody saying somebody was dying. I said Dr. Roth would be right over, but before I could even tell him, the phone started ringing again. And the minute I hung up the receiver, it rang again. That was the beginning of a terrible night. From that minute on, the phone never stopped ringing. That's the honest truth. And they were all alike. Everybody who called up said the same thing. Pain in the abdomen. Splitting headache. Nausea and vomiting. Choking and couldn't get their breath. Coughing up blood. But as soon as I got over my surprise, I calmed down. Hysterical people always end up by making me feel calm. Anyway, I managed to make a list of the first few calls and gave it to Dr. Roth. He was standing there with his hat and coat on and his bag in his hand and chewing on his cigar, and he took the list and shook his head and went out. Then I called Dr. Koehler, but his line was busy. I don't remember much about the next hour. All I know is I kept trying to reach Dr. Koehler and my phone kept ringing and my list of calls kept getting longer and longer."

One of the calls that lengthened Miss Stack's list was a summons to the home of August Z. Chambon, the burgess, or mayor, of Donora. The patient was the Burgess's mother, a widow of seventy-four, who lives with her son and his wife. "Mother Chambon was home alone that afternoon," her daughter-in-law says. "August was in Pittsburgh on business and I'd gone downstreet to do some shopping. It took me forever, the fog was so bad. Even the inside of the stores was smoky. So I didn't get home until around five-thirty. Well, I opened the door and stepped into the hall, and there was Mother Chambon. She was lying on the floor, with her coat on and a bag of cookies spilled all over beside her. Her face was blue, and she was just gasping for breath and in terrible pain. She told me she'd gone around the corner to the bakery a few minutes before, and on the way back the fog had got her. She said she barely made it to the house. Mother Chambon has bronchial trouble, but I'd never seen her so bad before. Oh, I was frightened! I helped her up—I don't know how I ever did it—and got her into bed. Then I called the doctor. It took me a long time to reach his office, and then he wasn't in. He was out making calls. I was afraid to wait until he could get here—Mother Chambon was so bad, and at her age and all—so I called another doctor. He was out, too. Finally, I got hold of Dr. Levin and he said he'd come right over, and he finally did. He gave her an injection that made her breathe easier and something to put her to sleep. She slept for sixteen solid hours. But before Dr. Levin left, I told him that there seemed to be an awful lot of sickness going on all of a sudden. I was coughing a little myself. I asked him what was happening. 'I don't know,' he said. 'Something's coming off, but I don't know what.'"

Dr. Roth returned to his office at a little past six to replenish his supply of drugs. By then, he, like Dr. Levin, was aware that something was coming off. "I knew that whatever it was we were up against was serious," he says. "I'd seen some very pitiful cases, and they weren't all asthmatics or chronics of any kind. Some were people who had never been bothered by fog before. I was worried, but I wasn't bewildered. It was no mystery. It was obvious—all the symptoms pointed to it—that the fog and smoke were to blame. I didn't think any further than that. As a matter of fact, I didn't have time to think or wonder. I was too damn busy. My biggest problem was just getting around. It was almost impossible to drive. I even had trouble finding the office. McKean Avenue was solid coal smoke. I could taste the soot when I got out of the car, and my chest felt tight, On the way up the stairs, I started coughing and I couldn't stop. I kept coughing and choking until my stomach turned over. Fortunately, Helen was out getting something to eat —I just made it to the office and into the lavatory in time. My God, I was sick! After a while, I dragged myself into my office and gave myself an injection of adrenalin and lay back in a chair. I began to feel better. I felt so much better I got out a cigar and lighted up. That practically finished me. I took one pull, and went into another paroxysm of coughing. I probably should have known better—cigars had tasted terrible all day—but I hadn't had that reaction before. Then I heard the phone ringing. I guess it must have been ringing off and on all along. I thought about answering it, but I didn't have the strength to move. I just lay there in my chair and let it ring."

When Miss Stack came into the office a few minutes later, the telephone was still ringing. She had answered it and added the call to her list before she realized that she was not alone. "I heard someone groaning," she says. "Dr. Roth's door was open and I looked in. I almost jumped, I was so startled. He was slumped down in his chair, and his face was brick red and dripping with perspiration. I wanted to help him, but he said there wasn't anything to do. He told me what had happened. 'I'm all right now,' he said. 'I'll get going again in a minute. You go ahead and answer the phone.' It was ringing again. The next I knew, the office was full of patients, all of them coughing and groaning. I was about ready to break down and cry. I had talked to Dr. Koehler by that time and he knew what was happening. He had been out on calls from home. 'I'm coughing and sick myself,' he said, 'but I'll got out again as soon as I can.' I tried to keep calm, but with both Doctors sick and the office full of patients and the phone ringing, I just didn't know which way to turn. Dr. Roth saw two or three of the worst patients. Oh, he looked ghastly! He really looked worse than some of the patients. Finally, he said he couldn't see any more, that the emergency house calls had to come first, and grabbed up his stuff and went out. The office was still full of patients, and I went around explaining things to them. It was awful. There wasn't anything to do but close up, but I've never felt so heartless. Some of them were so sick and miserable. And right in the middle of everything the parade came marching down the street. People were cheering and yelling, and the bands were playing. I could hardly believe my ears. It just didn't seem possible."

The sounds of revelry that reached Miss Stack were deceptive. The parade, though well attended, was not an unqualified success. "I went out for a few minutes and watched it," the younger Mrs. Chambon says. "It went right by our house. August wasn't home yet, and after what had happened to Mother Chambon, I thought it might cheer me up a little. It did and it didn't. Everybody was talking about the fog and wondering when it would end, and some of them had heard there was sickness, but nobody seemed at all worried. As far as I could tell, all the sick people were old. That made things look not too bad. The fog always affects the old people. But as far as the parade was concerned, it was a waste of time. You really couldn't see a thing. They were just like shadows marching by. It was kind of uncanny. Especially since most of the people in the crowd had handkerchiefs tied over their nose and mouth to keep out the smoke. All the children did. But, even so, everybody was coughing. I was glad to get back in the house. I guess everybody was. The minute it was over, everybody scattered. They just vanished. In two minutes there wasn't a soul left on the street. It was as quiet as midnight."

Among the several organizations that participated in the parade was the Donora Fire Department. The force consists of about thirty volunteers and two full-time men. The latter, who live at the firehouse, are the chief, John Volk, a wiry man in his fifties, and his assistant and driver, a hard, round-faced young man named Russell Davis. Immediately after the parade, they returned to the firehouse. "As a rule," Chief Volk says, "I like a parade. We've got some nice equipment here, and I don't mind showing it off. But I didn't get much pleasure out of that one. Nobody could see us, hardly, and we couldn't see them. That fog was black as a derby hat. It had us all coughing. It was a relief to head for home. We hadn't much more than got back to the station, though, and got the trucks put away and said good night to the fellows than the phone rang. Russ and I were just sitting down to drink some coffee. I dreaded to answer it. On a night like that, a fire could have been real mean. But it wasn't any fire. It was a fellow up the street, and the fog had got him. He said he was choking to death and couldn't get a doctor, and what he wanted was our inhalator. He needed air. Russ says I just stood there with my mouth hanging open. I don't remember what I thought. I guess I was trying to think what to do as much as anything else. I didn't disbelieve him—he sounded half dead already—but, naturally, we're not supposed to go running around treating the sick. But what the hell, you can't let a man die! So I told him O.K. I told Russ to take the car and go. The way it turned out, I figure we did the right thing. I've never heard anybody say different."

"That guy was only the first," Davis says. "From then on, it was one emergency call after another. I didn't get to bed until Sunday. Neither did John. I don't know how many calls we had, but I do know this: We had around eight hundred cubic feet of oxygen on hand when I started out Friday night, and we ended up by borrowing from McKeesport and Monessen and Monongahela and Charleroi and everywhere around here. I never want to go through a thing like that again. I was laid up for a week after. There never was such a fog. You couldn't see your hand in front of your face, day or night. Hell, even inside the station the air was blue. I drove on the left side of the street with my head out the window, steering by scraping the curb. We've had bad fogs here before. A guy lost his car in one. He'd come to a fork in the road and didn't know where he was, and got out to try and tell which way to go. When he turned back to his car, he couldn't find it. He had no idea where it was until, finally, he stopped and listened and heard the engine. That guided him back. Well, by God, this fog was so bad you couldn't even get a car to idle. I'd taken my foot off the accelerator and—bango!—the engine would stall. There just wasn't any oxygen in the air. I don't know how I kept breathing. I don't know how anybody did. I found people laying in bed and laying on the floor. Some of them were laying there and they didn't give a damn whether they died or not. I found some down in the basement with the furnace draft open and their head stuck inside, trying to get air that way. What I did when I got to a place was throw a sheet or a blanket over the patient and stick a cylinder of oxygen underneath and crack the valves for fifteen minutes or so. By God, that rallied them! I didn't take any myself. What I did every time I came back to the station was have a little shot of whiskey. That seemed to help. It eased my throat. There was one funny thing about the whole thing. Nobody seemed to realize what was going on. Everybody seemed to think he was the only sick man in town. I don't know what they figured was keeping the doctors so busy. I guess everybody was so miserable they just didn't think."

Toward midnight, Dr. Roth abandoned his car and continued his rounds on foot. He found not only that walking was less of a strain but that he made better time. He walked the streets all night, but he was seldom lonely. Often, as he entered or left a house, he encountered a colleague. "We all had practically the same calls," Dr. M. J. Hannigan, the president of the Donora Medical Association, says. "Some people called every doctor in town. It was pretty discouraging to finally get someplace and drag yourself up the steps and then be told that Dr. So-and-So had just been there. Not that I blame them, though. Far from it. There were a couple of times when I was about ready to call for help myself. Frankly, I don't know how any of us doctors managed to hold out and keep going that night."

Not all of them did. Dr. Koehler made his last call that night at one o'clock. "I had to go home," he says. "God knows I didn't want to. I'd hardly made a dent in my list. Every time I called home or the Physicians' Exchange, it doubled. But my heart gave out. I couldn't go on any longer without some rest. The last thing I heard as I got into bed was my wife answering the phone. And the phone was the first thing I heard in the morning. It was as though I hadn't been to sleep at all." While Dr. Koehler was bolting a cup of coffee, the telephone rang again. This time, it was Miss Stack. They conferred briefly about the patients he had seen during the night and those he planned to see that morning. Among the latter was a sixty-four-year-old steelworker named Ignatz Hollowitti. "One of the Hollowitti girls, Dorothy, is a good friend of mine," Miss Stack says. "So as soon as I finished talking to Dr. Koehler, I called her to tell her that Doctor would be right over. I wanted to relieve her mind. Dorothy was crying when she answered the phone. I'll never forget what she said. She said, 'Oh, Helen—my dad just died! He's dead!' I don't remember what I said. I was simply stunned. I suppose I said what people say. I must have. But all I could think was, My gosh, if people are dying—why, this is tragic! Nothing like this has ever happened before!"

Mr. Hollowitti was not the first victim of the fog. He was the sixth. The first was a retired steelworker of seventy named Ivan Ceh. According to the records of the undertaker who was called in—Rudolph Schwerha, whose establishment is the largest in Donora—Mr. Ceh died at one-thirty Saturday morning. "I was notified at two," Mr. Schwerha says. "There is a note to such effect in my book. I thought nothing, of course. The call awakened me from sleep, but in my profession anything is to be expected. I reassured the bereaved and called my driver and sent him for the body. He was gone forever. The fog that night was impossible. It was a neighborhood case—only two blocks to go, and my driver works quick—but it was thirty minutes by the clock before I heard the service car in the drive. At that moment, again the phone rang. Another case. Now I was surprised. Two different cases so soon together in this size town doesn't happen every day. But there was no time then for thinking. There was work to do. I must go with my driver for the second body. It was in the Sunnyside section, north of town, too far in such weather for one man alone. The fog, when we got down by the mills, was unbelievable. Nothing could be seen. It was like a blanket. Our fog lights were useless, and even with the fog spotlight on, the white line in the street was invisible. I began to worry. What if we should bump a parked car? What if we should fall off the road? Finally, I told my driver, 'Stop! I'll take the wheel. You walk in front and show the way.' So we did that for two miles. Then we were in the country. I know that section like my hand, but we had missed the house. So we had to turn around and go back. That was an awful time. We were on the side of a hill, with a terrible drop on one side and no fence. I was afraid every minute. But we made it, moving by inches, and pretty soon I found the house. The case was an old man and he had died all of a sudden. Acute cardiac dilation. When we were ready, we started back. Then I began to feel sick. The fog was getting me. There was an awful tickle in my throat. I was coughing and ready to vomit. I called to my driver that I had to stop and get out. He was ready to stop, I guess. Already he had walked four or five miles. But I envied him. He was well and I was awful sick. I leaned against the car, coughing and gagging, and at last I riffled a few times. Then I was much better. I could drive. So we went on, and finally we were home. My wife was standing at the door. Before she spoke, I knew what she would say. I thought, Oh, my God —another! I knew it by her face. And after that came another. Then another. There seemed to be no end. By ten o'clock in the morning, I had nine bodies waiting here. Then I heard that De Rienzo and Lawson, the other morticians, each had one. Eleven people dead! My driver and I kept looking at each other. What was happening? We didn't know. I thought probably the fog was the reason. It had the smell of poison. But we didn't know."

Mr. Schwerha's bewilderment was not widely shared. Most Donorans were still unaware Saturday morning that anything was happening. They had no way of knowing. Donora has no radio station, and its one newspaper, the *Herald-American,* is published only five days a week, Monday through Friday. It was past noon before a rumor of widespread illness began to drift through town. The news reached August Chambon at about two o'clock. In addition to being burgess, an office that is more an honor than a livelihood, Mr. Chambon operates a moving-and-storage business, and he had been out of town on a job all morning. "There was a message waiting for me when I got home," he says. "John Elco, of the Legion, had called and wanted me at the Borough Building right away. I wondered what the hell, but I went right over. It isn't like John to get excited over nothing. The fog didn't even enter my mind. Of course, I'd heard there were some people sick from it. My wife had told me that. But I hadn't paid it any special significance. I just thought they were like Mother—old people that were always bothered by fog. Jesus, in a town like this you've got to expect fog. It's natural. At least, that's what I thought then. So I was astonished when John told me that the fog was causing sickness all over town. I was just about floored. That's a fact. Because I felt fine myself. I was hardly even coughing much. Well, as soon as I'd talked to John and the other fellows he had rounded up, I started in to do what I could. Something had already been done. John and Cora Vernon, the Red Cross director, were setting up an emergency-aid station in the Community Center. We don't have a hospital here. The nearest one is at Charleroi. Mrs. Vernon was getting a doctor she knew there to come over and take charge of the station, and the Legion was arranging for cars and volunteer nurses. The idea was to get a little organization in things—everything was confused as hell—and also to give our doctors a rest. They'd been working steady for thirty-six hours or more. Mrs. Vernon was fixing it so when somebody called a doctor's number, they would be switched to the Center and everything would be handled from there. I've worked in the mills and I've dug coal, but I never worked any harder than I worked that day. Or was so worried. Mostly I was on the phone. I called every town around here to send supplies for the station and oxygen for the firemen. I even called Pittsburgh. Maybe I overdid it. There was stuff pouring in here for a week. But what I wanted to be was prepared for anything. The way that fog looked that day, it wasn't ever going to lift. And then the rumors started going around that now people were dying. Oh, Jesus! Then I was scared. I heard all kinds of reports. Four dead. Ten dead. Thirteen dead. I did the only thing I could think of. I notified the State Health Department, and I called a special meeting of the Council and our Board of Health and the mill officials for the first thing Sunday morning. I wanted to have it right then, but I couldn't get hold of everybody—it was Saturday night. Every time I looked up from the phone, I'd hear a new rumor. Usually a bigger one. I guess I heard everything but the truth. What I was really afraid of was that they might set off a panic. That's what I kept dreading. I needn't have worried, though. The way it turned out, half the town had hardly heard that there was anybody even sick until Sunday night, when Walter Winchell opened his big mouth on the radio. By then, thank God, it was all over."

The emergency-aid station, generously staffed and abundantly supplied with drugs and oxygen inhalators, opened at eight o'clock Saturday night. "We were ready for anything and prepared for the worst," Mrs. Vernon says. "We even had an ambulance at our disposal. Phillip DeRienzo, the undertaker, loaned it to us. But almost nothing happened. Altogether, we brought in just eight patients. Seven, to be exact. One was dead when the car arrived. Three were very bad and we sent them to the hospital in Charleroi. The others we just treated and sent home. It was really very queer. The fog was as black and nasty as ever that night, or worse, but all of a sudden the calls for a doctor just seemed to trickle out and stop. It was as though everybody was sick who was going to be sick. I don't believe we had a call after midnight. I knew then that we'd seen the worst of it."

Dr. Roth had reached that conclusion, though on more slender evidence, several hours before. "I'd had a call about noon from a woman who said two men roomers in her house were in bad shape," he says. "It was nine or nine-thirty by the time I finally got around to seeing them. Only, I never saw them. The landlady yelled up to them that I was there, and they yelled right back, 'Tell him never mind. We're O.K. now.' Well, that was good enough for me. I decided things must be letting up. I picked up my grip and walked home and fell into bed. I was dead-beat."

There was no visible indication that the fog was beginning to relax its smothering grip when the group summoned by Burgess Chambon assembled at the Borough Building the next morning to discuss the calamity. It was another soggy, silent, midnight day. "That morning was the worst," the Burgess says. "It wasn't just that the fog was still hanging on. We'd begun to get some true facts. We didn't have any real idea how many people were sick. That didn't come out for months. We thought a few hundred. But we did have the number of deaths. It took the heart out of you. The rumors hadn't come close to it. It was eighteen. I guess we talked about that first. Then the question of the mills came up. The smoke. L. J. Westhaver, who was general superintendent of the steel and wire works then, was there, and so was the head of the zinc plant, M. M. Neale. I asked them to shut down for the duration. They said they already had. They had started banking the fires at six that morning. They went on to say, though, that they were sure the mills had nothing to do with the trouble. We didn't know what to think. Everybody was at a loss to point the finger at anything in particular. There just didn't seem to be any explanation. We had another meeting that afternoon. It was the same thing all over again. We talked and we wondered and we worried. We couldn't think of anything to do that hadn't already been done. I think we heard about the nineteenth death before we broke up. We thought for a week that was the last. Then one more finally died. I don't remember exactly what all we did or said that afternoon. What I remember is after we broke up. When we came out of the building, it was raining. Maybe it was only drizzling then—I guess the real rain didn't set in until evening—but, even so, there was a hell of a difference. The air was different. It didn't get you any more. You could breathe."

The investigation of the disaster lasted almost a year. It was not only the world's first full-blooded examination of the general problem of air pollution but one of the most exhaustive inquiries of any kind ever made in the field of public health. Its course was directed jointly by Dr. Joseph Shilen, director of the Bureau of Industrial Hygiene of the Pennsylvania Department of Health, and Dr. J. G. Townsend, chief of the Division of Industrial Hygiene of the United States Public Health Service, and at times it involved the entire technical personnel of both agencies. The Public Health Service assigned to the case nine engineers, seven physicians, six nurses, five chemists, three statisticians, two meteorologists, two dentists, and a veterinarian. The force under the immediate direction of Dr. Shilen, though necessarily somewhat smaller, was similarly composed.

The investigation followed three main lines, embracing the clinical, the environmental, and the meteorological aspects of the occurrence. Of these, the meteorological inquiry was the most nearly conclusive. It was also the most reassuring. It indicated that while the situation of Donora is unwholesomely conducive to the accumulation of smoke and fog, the immediate cause of the October, 1948, visitation was a freak of nature known to meteorologists as a temperature inversion. This phenomenon is, as its name suggests, characterized by a temporary, and usually brief, reversal of the normal atmospheric conditions, in which the air near the earth is warmer than the air higher up. Its result is a more or less complete immobilization of the convection currents in the lower air by which gases and fumes are ordinarily carried upward, away from the earth.

The clinical findings, with one or two exceptions, were more confirmatory than illuminating. One of the revelations, which was gleaned from several months of tireless interviewing, was that thousands, rather than just hundreds, had been ill during the fog. For the most part, the findings demonstrated, to the surprise of neither the investigators nor the Donora physicians, that the affection was essentially an irritation of the respiratory tract, that its severity increased in proportion to the age of the victim and his predisposition to cardio-respiratory ailments, and that the ultimate cause of death was suffocation.

The environmental study, the major phase of which was an analysis of the multiplicity of gases emitted by the mills, boats, and trains, was, in a positive sense, almost wholly unrewarding. It failed to determine the direct causative agent. Still, its results, though negative, were not without value. They showed, contrary to expectation, that no one of the several stack gases known to be irritant—among them fluoride, chloride, hydrogen sulphide, cadmium oxide, and sulphur dioxide—could have been present in the air in sufficient concentration to produce serious illness. "It seems reasonable to state," Dr. Helmuth H. Schrenk, chief of the Environmental Investigations Branch of the Public Health Service's Division of Industrial Hygiene, has written of this phase of the inquiry, "that while no single substance was responsible for the . . . episode, the syndrome could have been produced by a combination, or summation of the action, of two or more of the contaminants. Sulphur dioxide and its oxidation products, together with particulate matter [soot and fly ash], are considered significant contaminants. However, the significance of the other irritants as important adjuvants to the biological effects cannot be finally estimated on the basis of present knowledge. It is important to emphasize that information available on the toxicological effects of mixed irritant gases is meagre and data on possible enhanced action due to absorption of gases on particulate matter is limited." To this, Dr. Leonard A. Scheele, Surgeon General of the Service, has added, "One of the most important results of the study is to show us what we do not know."

Funeral services for most of the victims of the fog were held on Tuesday, November 2nd. Monday had been a day of battering rain, but the weather cleared in the night, and Tuesday was fine. "It was like a day in spring," Mr. Schwerha says. "I think I have never seen such a beautiful blue sky or such a shining sun or such pretty white clouds. Even the trees in the cemetery seemed to have color. I kept looking up all day."

Senator Gaylord Nelson (1916-   ) of Wisconsin, the author of the following essay, is one of the country's most active and outspoken critics of the destruction of the American environment. After his introductory remarks, Senator Nelson, who has introduced much legislation to protect and restore the quality of our environment, concentrates his discussion on the pollution of the nation's waters. Because he comes from a Great Lakes state and one which has in the past prided itself on its clean and sparkling lakes and streams, Senator Nelson conveys a sense of personal loss as he catalogues the dead and dying waterways of the Great Lakes region: the Ohio River, the Maumee, the Cuyahoga, Green Bay, Lake Erie, and others. After citing the primary sources of water pollution and the enormous dollar costs of cleaning up, Senator Nelson can nevertheless find some encouragement in the public's willingness to pay the price to save their waters. The public, he says, is far ahead of its officials in its awareness of this crisis. He further argues that the largest public body, the national government, must step in to assume the major financial responsibility for the cleanup.

THE NATIONAL POLLUTION SCANDAL

by Gaylord Nelson

The natural environment of America—the woods and waters and wildlife, the clear air and blue sky, the fertile soil and the scenic landscape—is threatened with destruction. Our growing population and expanding industries, the explosion of scientific knowledge, the vast increase in income levels, leisure time, and mobility—all of these powerful trends are exerting such pressure on our natural resources that many of them could be effectively ruined over the next ten or fifteen years. Our overcrowded parks are becoming slums. Our birds and wildlife are being driven away or killed outright. Scenic rural areas are blighted by junkyards and billboards, and neon blight soils the outskirts of most cities. In our orgy of expansion, we are bulldozing away the natural landscape and building a cold new world of concrete and aluminum. Strip miners' shovels are tearing away whole mountains and spreading ugly wastes for miles around. America the affluent is well on the way to destroying America the beautiful. Of all these developments, the most tragic and the most costly is the rapidly mounting pollution of our lakes and streams.

Perhaps the pain is more intense for a Senator from a state like Wisconsin, bordered on three sides by the Great Lakes and the Mississippi, blessed with 8,000 inland lakes and hundreds of rivers and trout streams. Actually,

"The National Pollution Scandal" From the February 1967 issue of *The Progressive*. Reprinted by permission of the author and the publisher.

our state seems rather fortunate at the moment. A yachtsman on Lake Superior can raise a bucket of water still crystal-clear and cold enough to drink with delight. Canoeists on the St. Croix or Wolf Rivers still shoot through frothing rapids of sparkling water, and catch fish in the deep, swirling pools.

But the bell is tolling for Wisconsin just as for all the nation. A recent survey of twelve major river basins in southeastern Wisconsin found not a single one fit even for the partial body contact involved in fishing or wading. A competent governmental agency concluded that 754 miles of rivers in this region had been turned into open sewers. Beaches along Lake Michigan, a vast blue sea with seemingly limitless quantities of fresh water, are being closed to swimmers. A sordid ocean of pollution is pouring into the Mississippi from the Minneapolis-St. Paul urban complex. The first serious signs of pollution are soiling Lake Superior, and our small inland lakes are, one by one, becoming murky and smelly and choked with algae.

Elsewhere, all across the nation, the same tragedy is being enacted, although in many areas the curtain already has come down. The waters are already ruined. Every major river system in America is seriously polluted, from the Androscoggin in Maine to the Columbia in the far Northwest. The rivers once celebrated in poetry and song—the Monongahela, the Cumberland, the Ohio, the Hudson, the Delaware, the Rio Grande—have been blackened with sewage, chemicals, oil, and trash. They are sewers of filth and disease. The Monongahela, which drains the mining and industrial areas of West Virginia and Pennsylvania, empties the equivalent of 200,000 tons of sulfuric acid each year into the Ohio River—which in turn is the water supply for millions of people who use and re-use Ohio River water many times over.

National attention has been centered on once beautiful Lake Erie, the great lake which is the recreational front yard of Buffalo, Cleveland, Toledo and Detroit, and which supplies water for ten million Americans. A Public Health Service survey of Lake Erie made the shocking discovery that, in the 2,600 square mile heart of the lake, there was no dissolved oxygen at all in the water. The lake in this vast area could support no desirable aquatic life, only lowly creatures such as bloodworms, sludgeworms, sowbugs, and bloodsuckers.

Along with the germs and industrial acids which pour into Lake Erie are millions of pounds of phosphates, a major ingredient in detergents. Each pound of phosphate will propagate 700 pounds of algae. Beneath the waters of this great lake, largely hidden from sight, a hideous cancer-like growth of algae is forming. As algae blooms and dies, it becomes a pollutant itself. It robs the lake of still more oxygen—and it releases the phosphate to grow another crop of algae. Lake Erie is a product of its tributaries. A Public Health Service study of these American sewers is horrifying to read.

The Maumee River flows from Fort Wayne, Indiana, through Defiance and Napoleon, Ohio, and on to Toledo, where it joins the lake. Even as far upstream as Fort Wayne, the river has insufficient oxygen to support anything but trash fish and lower organisms, and as it flows toward Lake Erie conditions get steadily worse. The count of coliform bacteria runs as high as 24,000 times the allowable maximum under Federal drinking water standards. The concentration of carbolic acid, a byproduct of steelmaking, runs up to 137 times the allowable maximum. A packing company dumps 136 pounds of oil per day into the Maumee River. A plating company dumps thirty-eight pounds of cyanide per day. Defiance, Ohio, closes its sewage plant entirely for one or two months each year, and all its raw sewage goes directly into the Maumee. Below Defiance, a foundry dumps cinders and ashes into the river. The Maumee is joined by the Auglaize River, which is even more polluted than the Maumee, and is especially rich in ammonia compounds. At Napoleon, Ohio, the city draws its drinking water from the sordid Maumee, and a soup company draws off ten million gallons a day for soup processing. (The firm assures me that its modern water treatment plant, complete with carbon filters, can "polish the water to a high quality.") Below Napoleon, things get really bad. Forty per cent of samples taken by the Public Health Service showed presence of salmonella, an intestinal bacterium that can cause severe illness. As the Maumee flows into Lake Erie at Toledo, it gets its final dose of pollution—the effluent from the Toledo sewage plant and what the Public Health Service describes as "oil, scum, metallic deposits, and toxic materials."

Another Lake Erie tributary—the Cuyahoga—which flows into the lake at Cleveland, is described by the Public Health Service as "debris-filled, oil-slicked, and dirty-looking throughout." It is loaded with coliform bacteria and salmonella. It is so polluted with oil that it frequently catches fire. Structures known as "fire breaks" have been built out into the river to fight these blazes. In the Cleveland harbor, the Public Health Service could find virtually no conventional aquatic life. However, the sludgeworms which thrive on organic matter were well represented—400,000 per square meter on the harbor bottom.

That is the story of Lake Erie, and although it is so shocking and disgusting as to deserve urgent national attention, it is not unique. Southern Lake Michigan, ringed with oil refineries, steel mills, and municipal sewage outfalls, may be even worse. Scientists estimate that it would take 100 years to replace the polluted water of southern Lake Michigan, and some consider the pollution in this area irreversible.

We have our own Wisconsin pollution scandal in Green Bay, a magnificent recreational body of water in northeastern Wisconsin, widely known as a yachtsman's paradise and site of a multimillion dollar resort industry. This "Cape Cod of Wisconsin" is threatened with ruin by a tide of pollution which is moving up the bay at the rate of more than one mile per year. The pollution comes from such rivers as the Fox, the Peshtigo, the Oconto, and the Menominee, which drain large areas of Wisconsin and northern Michigan.

The experience in Lake Erie, Lake Michigan, and Green Bay has convinced many experts of this chilling fact: It is a definite possibility that the Great Lakes—the greatest single source of fresh water in the world—could be effectively destroyed by pollution in the years ahead. If this were to happen, it would be the greatest natural resource disaster in modern history.

That is the outline of this new American tragedy. The obvious question now is, what can be done about it? First, I think we must learn what a complex and widespread problem we face in water pollution. Like crime, like death on the highway, pollution is a social problem which extends throughout our society. There is no single villain, and there is no simple answer. It must be attacked for what it is—a sinister byproduct of the prosperous, urbanized, industrialized world in which we live. We must take care not to ride off in pursuit of just one villain—such as city sewage, or industrial waste, or detergents, or toilet wastes from boats; this is a battle which must be fought with skill and courage on many different fronts. Nor should we be fooled by the strategy of many polluters, who argue, in effect: "The pollution which we cause is minor compared to the big, nation-wide problem. Why not leave us alone and go after the big offenders?" Even some of the lesser offenders in the pollution crisis could ruin us in time.

The primary sources of pollution are these:

MUNICIPAL SEWAGE Despite heroic efforts and heavy investments by many cities, our municipal sewage treatment plants are woefully inadequate. Some cities have no treatment at all; others remove only part of the pollutants found in sewage. As a result, the effluent discharged by our cities today (treated and untreated) is equivalent to the *untreated* sewage from a nation of seventy-five million people.

INDUSTRIAL POLLUTION is roughly twice as big a problem as municipal sewage. Despite tremendous investments in research and treatment plant construction by some industries, the overall record is terrible. Some industries feel they cannot remain competitive if they spend heavily for treatment plants. Communities and states are reluctant to push them too far. As a result, industrial wastes (treated and untreated) now discharged into our waters are presently equal to the *untreated* sewage of a nation of 165 million people.

SEPTIC TANKS Vast sections of the nation have no sewer collection or treatment system at all. In such areas, underground septic tanks, often poorly made and undersized, are expected to distribute wastes into the soil. They overflow into natural watercourses, they leak bacteria and detergents into underground wells, and they are destroying lakes by filling them with nutrients that foster heavy growths of algae.

SHIPS AND MARINE TERMINALS In selected areas, the discharge of toilet wastes, oil, garbage, and rubbish from ships and shoreline installations is a major problem. For some reason, this form of pollution is widely tolerated and enforcement of laws forbidding it is virtually nonexistent.

PESTICIDES The terrifying prospect of spreading poison all over the globe confronts us. We now use more than 700 million pounds a year of synthetic pesticides and agricultural chemicals of 45,000 varieties. This volume is expected to increase tenfold in the next twenty years. Many of these poisons persist forever in the environment, and their concentration builds up geometrically as they progress through the food chain (water, seaweed, fish, birds, mammals). DDT residue has been discovered in penguins in Antarctica, in reindeer in Alaska, in seals, and in fish caught in remote areas of the Pacific Ocean. One part of DDT in one billion parts of water will kill blue crabs in eight days.

SILT One of the most serious pollutants all over the world is the dirt which washes into our waters from off the land. This somewhat natural problem is disastrously aggravated by contemporary trends—widespread clearing of land for subdivisions and shopping centers; construction of highways and parking lots (which cause rapid runoff) and the intensive development of lakeshores and riverbanks. Controlling surface runoff and the siltation which it causes is complicated by our patchwork of political boundaries and the lack of coordinated government planning.

DETERGENTS, FERTILIZERS, AND OTHER CHEMICALS Some of these commonly used substances pass through even good waste treatment systems and become persistent pollutants. Such pollution can be eliminated only by changing the composition of such substances, regulating their use, or devising new removal techniques.

Obviously, any nationwide problem made up of so many elements is extremely difficult to attack. Yet I believe that the rapidly accelerating destruction of our natural resources is our number one domestic problem, and the greatest of all our resource problems is water pollution. If we are to meet this pollution threat, if we are to save the waters of America and preserve this most indispensable part of our natural environment, we must make the war on pollution a high priority matter at every level of government—local, state and Federal—and we must insist that private industry do likewise. Baffling and complicated as the pollution problem is, it is not insoluble. There is no reason in the world why a great and prosperous nation, with the money and know-how to shoot man to the moon, cannot prevent its lakes and rivers from being destroyed and its life-giving water supplies endangered. Just as there is no single cause of pollution, so is there no single solution to the problem.

Consider the question of what to do about municipal sewage and industrial wastes. Why do we tolerate a situation where these two sources alone pour into our waters each year the equivalent of the completely untreated sewage of a nation of 240 million persons? Here it is largely a matter of lack of money, aggravated in some cases by a shocking lack of public concern. There are now more than 1300 communities which have sewer systems but discharge their wastes into the waters without any treatment at all. These communities have a population of more than eleven million people. How such a condition could exist in the year 1966—when it is generally illegal to throw a gum wrapper out of a car window—is inconceivable.

We have another 1300 communities—with almost seventeen million population—which treat their wastes but in a completely inadequate manner. In most cases, these are communities which use what is known as "primary" treatment. They screen their sewage and let the solids settle out, but they do not remove dissolved solids, salts, chemicals, bacteria, and special problems such as detergents. Every community should have what is known as "secondary" treatment, under which sewage—after primary treatment—is held in holding tanks, brought into contact with air and biologically active sludge, so that bacteria have a chance to consume the pollutants.

The Conference of State Sanitary Engineers estimates that it would cost $1.8 billion to provide adequate sewage collection and treatment for these communities which now have no treatment or completely inadequate treatment. But even this would still leave us with a massive municipal pollution problem. Even good secondary treatment removes only eighty per cent to ninety per cent of the pollutants. Chicago, for instance, with a good secondary treatment plant, discharges treated effluent which is equivalent to the untreated, raw sewage of one million people. It dumps 1,800 tons of solids per day into the Illinois waterway. At the rate the pollution load is increasing it is estimated that even if all communities have secondary treatment plants by 1980, the total amounts of pollutants reaching watercourses would still be the same as today. Obviously, we need a massive program to build highly effective city sewage treatment plants.

It is also obvious that local property taxes cannot support such a gigantic investment, and that if we wait for communities to do this on their own, it will never be done. Most state budgets also are severely strained, so much of this burden is going to have to be borne by the Federal government—if we want the job done early enough to be effective. The Senate Air and Water Pollution subcommittee estimates that it will cost $20 billion to provide secondary treatment in plants serving eighty per cent of the population and more advanced treatment in plants serving the other twenty per cent. We have had a Federal program to assist communities in building such treatment plants for the past ten years, but it has been inadequate. It has recently been greatly improved, but it is still inadequate. In the past it has provided grants of up to thirty per cent within the limits of available funds. The most recent act—the Clean Waters Restoration Act of 1966—authorizes a total of about $3.6 billion over the next five years ($150 million in 1967, $450 million in 1968, $700 million in 1969, $1 billion in 1970, and $1.25 billion in 1971). A community can get a grant for up to fifty per cent of the cost of a project, provided the state pays twenty-five per cent and provided water quality standards have been established.

New York needs an estimated $1.7 billion for new sewage plants. The new law would give it a total of only $307 million. Ohio needs $1 billion and would get $180 million. Wisconsin needs $286 million and would get $75 million. If we are serious about the Federal government paying fifty per cent of the cost of eliminating municipal pollution, then Washington must provide $10 billion—not $3.6 billion—and even then we will be expecting our hard-pressed states and communities to come up with another $10 billion. Personally, I think it is unrealistic to expect the states and localities to assume a burden of this size. And I do not think the nation can sit by and wait while its communities struggle to build up the financial resources and the political courage needed to do the job. I think we should get sewage treatment plants built the way we are getting interstate highways built—by offering ninety per cent Federal financing. I have introduced legislation which would establish such a program.

The municipal sewage problem is complicated by another problem—combined storm and sanitary sewers. By combining storm water and human wastes in one sewer system, many cities build up such a tremendous load during rainstorms that their sewage treatment plants cannot handle it. They have had to install automatic devices which divert the combined sewer load directly into lakes or streams whenever it gets above a certain level. In this manner, sixty-five billion gallons of raw, untreated sewage goes into our lakes and rivers each year. Most cities are separating storm and sanitary sewers in new subdivisions, but the task of separating the sewers in the older areas is a staggering one. Complete separation would cost an estimated $30 billion. It would cost $160 per resident in Washington, D. C., $215 in Milwaukee, $280 in Concord, New Hampshire. It would cost Wisconsin an estimated $186 million, Indiana $496 million, Michigan $970 million, New York and Illinois about $1.12 billion each. These are only general estimates of the direct costs and they do not take into account the disruption of traffic and the local economy caused by ripping up miles of underground sewers. In the hope of avoiding such costs, the Federal government has underwritten several research projects to see if this problem cannot be met in some other way—through temporary underground storage of sewer overflows, for instance, or by building smaller sanitary sewer pipes inside existing storm sewers.

The staggering problem of *industrial* pollution is virtually untouched today by our Federal anti-pollution programs, even though industry contributes twice as much pollution to our waters as do municipalities. If we do not step up our industrial waste treatment plant construction, the pollution effect of industrial wastes alone by 1970 will be equal to the untreated, raw sewage from our entire population. Industries are widely criticized for dumping wastes into our waters, and this criticism is often justified. They are pressured by local, state, and Federal officials. But some industries are able to avoid a serious crackdown against them by threatening to move. Most industries argue—sometimes effectively—that they cannot be expected to make massive investments in treatment plants if their competitors—often in different parts of the country—are not forced to do so.

I have come to the conclusion that the threat of enforcement alone is not going to solve our industrial pollution problem. We must provide direct financial assistance to see to it that the plants are built. I have introduced legislation to provide both loans and grants of up to fifty per cent to industries whose size and economic circumstances prevent them from assuming the full burden of providing their own facilities. I think such assistance should be carefully limited and should be for a short period, but I do not think we can avoid it. We are going to pay the cost of industrial pollution in one way or another—in the cost of the manufactured product, in taxes, or in ruined water resources.

But massive construction programs alone are not going to solve our municipal and industrial pollution problems. We need a tremendous expansion of Federally supported research to find completely new answers. Our whole waste disposal system, from the household toilet to the municipal sewage treatment plant, is a holdover from another era. The system should be studied and redesigned, using the latest scientific techniques, and fitted into a coordinated, nationwide system of waste disposal. Research grants should be made to private industry and universities to develop new methods and devices to refine, use, neutralize, or destroy pollutants. We should compute what our present waste disposal systems are costing us—including the loss in natural resources destroyed—and what alternative systems would cost.

Compared with municipal and industrial pollution, the other pollution problems I have mentioned are statistically small. For that reason, they are often ignored. But we cannot safely do that. Even if we managed to contain the flood of municipal and industrial pollution, the other sources could do fatal damage to our environment. Septic tanks must be controlled at the state and local level, and in many areas I think we must forbid new installations and work to replace existing ones with sewer systems. For instance, once an inland lake is ringed with cottages with septic tanks, it is doomed. Septic tanks must drain somewhere and in most lakeshore settings the natural drainage flow is into the lake. At the very least, this drainage will fertilize the lake, cause the rapid growth of algae, and turn the lake into a murky, foul-smelling mess.

Ship pollution is certainly serious enough to justify Federal action, even though such suggestions cause howls of protest from those who insist it "isn't practical." Why is it practical to install retention facilities on buses, house trailers, and aircraft but not on boats and ships? Obviously, we are willing to allow wastes to be dumped into our water supplies which we would never tolerate being dumped onto the land. We need Federal laws to require suitable facilities on all vessels using our navigable waters, and we need a better enforcement system to crack down on such disgraceful practices as dumping oil and pumping out oily ballast tanks on the Great Lakes and in our rivers.

The siltation problem can be controlled only through strict zoning and land use controls. We have got to prevent intensive development of our shorelines if we are to save our waters. Once a large portion of the natural vegetative cover is destroyed, the water resource is in danger. I believe that the Federal government should provide financial assistance to those willing to carry out soil conservation practices along our lakes and streams on a scale large enough to be meaningful.

Pesticides, detergents, and exotic new chemicals will plague us for years to come. New treatment systems may offer some hope for removing these substances, but I think they must be controlled directly. Those which cannot be removed safely in normal treatment processes, and those which have chemical structures which cause them to persist in our environment and to threaten fish, wildlife, and human health, should be banned or their use strictly regulated.

In speeches in some twenty-three states in the past four years, I have called for an emergency, crash program to fight water pollution. I have offered my estimate of the cost of conquering water pollution as $50 to $100 billion over the next decade. It now appears I may have been conservative. The Public Health Service now estimates that it will cost some $20 billion to clean up the Great Lakes alone, and the total national cost is now estimated at $100 billion. But everywhere I have gone I have found the public willing to pay this cost to save their waters. In fact, I think the public is far ahead of local, state, and Federal officials in facing up to this crisis. I think that citizens in most communities would support a sharp crackdown on local polluters of every variety. I think they want their states to establish high water quality standards, and then enforce them. I think they can be shown the need for bold regional action to deal with those vast interstate pollution problems (such as on the Mississippi and the Great Lakes) which obviously are too big for any community or any state to handle. And I think that the citizens of America now recognize that the destruction of the major river networks of the nation, the threatened destruction of the Great Lakes, and the slow ruination of our treasured inland lakes and trout streams is a calamity of such gigantic proportions as to deserve the urgent attention of all citizens and prompt action by the national government.

Gaylord Nelson's indictment of our nation's polluted waterways is expanded by Frank Graham, Jr. (1925-), in the following case history of a fish kill on the lower Mississippi River. Graham's essay is a description of the detective work which is necessary before the cause of a pollution kill can be established. Beyond that, Graham describes how this case remained unresolved even once the investigation had definitely traced the fish kill to wastes dumped into a tributary of the Mississippi at Memphis, Tennessee, by the Velsicol Chemical Corporation, manufacturers of the pesticide Endrin. The offending industry, backed by powerful allies including some within the federal government, fought the results of the investigation and opposed the demands for a cleanup. The belief that industry has a right to pollute is one of the last tenets of nineteenth-century robber baron capitalism; yet it is a conviction that dies hard, as Frank Graham shows here and as Barry Weisberg claims in "Alaska—The Ecology of Oil" (page 187).

Frank Graham, Jr., describes himself as not a scientist or an engineer but as a reporter with a strong interest in the preservation of the environment. He is the author of numerous articles and books on this subject. His most recent book is Since Silent Spring.

THE MISSISSIPPI FISH KILL

by Frank Graham, Jr.

On November 18, 1963, Robert LaFleur of Louisiana's Division of Water Pollution Control called the Public Health Service in Washington to report a massive fish kill in the lower Mississippi River and its tributary, the Atchafalaya. An estimated five million dead fish were floating, belly up, in the great muddy river which drains a third of the United States, provides drinking water for over a million southerners, and supports a vast segment of this country's fishing industry. LaFleur's telephone call, made reluctantly as are most pleas for help by state officials to the federal government, set in motion more than a routine investigation. Before many months had passed, roving teams of water "detectives" had touched off an uproar which was fueled by the fires from many of the basic moral, political, and business controversies of our time.

It was not the first such kill on the Mississippi in recent years. In the summer of 1958, fish, snakes, eels, and turtles died in enormous numbers in streams throughout the sugar-cane areas of southern Louisiana. State chemists could not detect toxic materials in either the water or the fish, but the carnage went on. In 1959 a state report noted that during that year's growing season "these kills reached alarming proportions with complaints being received by the Division of Water Pollution Control on an almost daily basis."

At least 30 major kills were reported during the summer of 1960. Kenneth Biglane, who was then chief of the Louisiana Water Pollution Division, recalls his department's frustration. "The sports public, indignant citizens, harassed elected officials and the press were all demanding that something be done. Dead fish were observed to be clogging the intake of the Franklin, Louisiana, power plant and were dying in Bayou Teche, a stream used as a source of drinking water for the town of Franklin."

In the fall of 1960, nearly four million fish died mysteriously in the Mississippi and the Atchafalaya. The dead fish included threadfin shad, fresh water drum, and buffalo, but by far the greatest number of victims were catfish (95 percent), the principal source of food for many of the poorer people in the river towns and bayou settlements. Although continued observation did not give scientists any clues to the cause of their death, all the dead and dying fish exhibited strikingly similar symptoms.

"Most of the catfish were bleeding about the mouth," Robert LaFleur recalls, "and many were bleeding about the fins. In every instance examination revealed that this was due to distention of the swim bladder and the digestive tract. The latter was devoid of food material and contained only gas and a small amount of bile-like frothy material. Analysis of the bottom organisms revealed that an abundant food supply was available. Dying fish were swimming at the surface, often inverted, in a very lethargic manner, and were easily captured by hand."

According to the state investigators, this was "the most spectacular fish mortality ever noted in Louisiana," but their only conclusion was that it might be attributed to "abdominal dropsy." When smaller but widespread kills occurred again in 1961 and 1962, the details were not reported to the United States Public Health Service because, according to Biglane, "We did not think pollution was responsible for the mortalities."

By 1963, however, Biglane had joined the Public Health Service in Washington, and he was among the federal pollution experts who were alarmed by the report of the massive fish kill reported that November. Louisiana officials, until then reluctant to call in the federal government on a problem which they considered rightfully belonged to the state, had thrown up their hands in despair. LaFleur's call to Washington was a plea for immediate help.

Teams of federal investigators descended on the lower Mississippi. They talked to local scientists, read reports on previous fish kills, and toured the stricken areas in small boats. There was no scarcity of evidence, and no scarcity of local people waiting to tell their stories. Harry McHugh, who lives along the Atchafalaya, told his in the Franklin *Banner-Tribune*.

"I can look out my front door," he wrote in January, 1964, "and see at least a few fish swimming crazily on the surface of the water—mostly shad and mullet. These will swim in such a fashion for a while, eventually losing their ability to swim, and then either drift on down with the tide toward the Gulf of Mexico or are washed up on the banks where they die."

A reporter for *The New Republic* learned that the streams and bayous were clogged with dead ducks, mostly fish-eating species like scaup and mergansers. "The bodies of turtles floated on the waters," *The New Republic* said. "Tough 150-pound garfish and catfish weighing 70 pounds surfaced, too weak to move. Crabs lay along the banks. Thousands of cranes and robins lay dead. The pencil-size white eels fishermen used for bait were scooped up dead by the net full. Alligators, once plentiful, have disappeared. Otter died or left the swamps."

Though many human beings lived on the product of those waters, there were no reports yet of illness and death among them. Economically, the story was different. The men who had fished the Cajun areas of Louisiana for years were hard hit. The wives of many were forced to go to work in the canneries, while the fishermen themselves took work in nearby towns or in the off-shore oil industry. "More fishermen have gone on jobs this year than ever before," one of them told a reporter from the *New York Times*.

"This time the poison hit all types of fish," another fisherman said. "So many died this winter that it got so I couldn't make $10 a week."

A wholesaler reported that he could buy only a small percentage of the fish usually available to him. "You can stand here on the dock any day of the fall and winter and see thousands of dead fish float by with the current," he said. "They would shoot up out of the water and just flop over. Many others died in the nets before the fishermen could bring them in."

These tales were confirmed by the experience of government scientists. Dr. Donald Mount, a young biologist with the Public Health Service, told of touring the Mississippi in an open powerboat near Baton Rouge.

"Thousands of catfish, drum, buffalo and shad were seen at the surface, unable to maintain an upright position and often having convulsions.

Literally acres of minnow schools were also in a state of hyperactivity and convulsions."

Dr. Mount also visited the delta areas where menhaden, a salt-water fish which enters estuaries to cast its eggs in late winter and early spring, were reported dying by the millions. Menhaden, in fact, made up by far the greatest proportion of the fish killed in the estuaries during the winter of 1963-64. Touring the harbor at Venice, 235 miles below Baton Rouge, Dr. Mount observed thousands of convulsing blue catfish, mullet, and menhaden. "On several occasions," he said, "we went into a dead-end canal, and as the boat approached the end, schools of menhaden were trapped and hundreds of these fish began convulsing, jumping onto the boat, the banks and against an oil drilling rig."

Scientists from various government agencies collected dead and dying fish, froze them, and shipped them to laboratories for study. Analysis by the U.S. Fish and Wildlife Service (Department of the Interior) proved that the fish were not killed by parasitic or bacterial disease. Botulism was ruled out, as were organic phosphorous insecticides and toxic concentrations of metals.

But government scientists were for the first time armed with recently developed techniques and machines which brought them within reach of a solution. At the Taft Sanitary Engineering Center in Cincinnati, investigators worked for weeks to find a clue to the mysterious fish kills. Tissues of dying catfish were fed to healthy catfish, but no viral or bacterial diseases were transmitted to them. A major breakthrough resulted, however, when scientists turned to the mud which had been dredged from affected areas of the river. Extracts of the tissues taken from dying fish were then dissolved in water where healthy fish swam. The healthy fish convulsed and died—symptoms exactly similar to those of the stricken fish in the Mississippi and Atchafalaya!

Teams of investigators (one from a private research laboratory) independently analyzed both the mud and the dead fish. Endrin, a chlorinated hydrocarbon insecticide, was found in every extract.

This deadly poison was not new to government scientists. For some years the Bureau of Commercial Fisheries (Department of the Interior) had been studying Endrin and similar poisons, Dieldrin and Aldrin, because of their effect on key commercial fish like shrimp and menhaden. These insecticides, marketed because of their extreme toxicity to arthropods (the phylum to which most insect pests belong), were naturally found to be especially harmful to shrimp, which are marine arthropods. Dr. C. M. Tarzwell, chief of the Public Health Service's Aquatic Biology Section, reports that, of all the substances he ever tested, Endrin is the most toxic to fish.

Endrin, indeed, was more than a laboratory goblin. It was known to Louisiana biologists as far back as 1958, the year of the first mysterious fish kill there. They noticed then a cause-and-effect relationship between the spraying of sugar-cane fields with Endrin (to control an insect called the sugar-cane borer) and the subsequent drainage of those fields after heavy rains. Thousands of fish immediately died. When the spraying stopped, the killing stopped. Kenneth Biglane, chief of the state's Division of Water Pollution Control at the time, clearly remembers that sequence of events.

"I have witnessed aerial applications of Endrin on sugar-cane fields around Houma, Thibodaux, and Donaldsonville, Louisiana," he says. "After subsequent rains, I have returned to such streams as Bayou Pierre Part near Donaldsonville, Bayou Black near Houma, and Bayou Chevrieul near Thibodaux, and I have seen thousands of fish and snakes, turtles, eels, dead and dying."

Yet all attempts to pin down Endrin as the cause of the fish kills during those years proved futile. The advanced techniques of fish autopsy and the sensitive instruments capable of detecting minute traces of foreign substances in water and mud (as low as a few parts per billion) were not then available to investigators. The now-familiar pattern of delay began to appear. The federal government was reluctant to move in because, as one Public Health Service scientist said, "it would destroy state relations and dry up future research." State officials, jealous of their sovereignty, refused to allow federal investigators to come in and perhaps put restrictions locally on the sale of suspected pesticides. The state, of course, could have taken strong measures of its own. In 1959, Louisiana limited itself to a public education program, pointing out the dangers of the careless use of Endrin.

"Regulations for the supervision of aerial applications," Biglane says, "the marking of cane fields, the plugging of drainage ditches leading away from fields, and the halting of applications immediately before a rain was forecast were recommended to the Louisiana Stream Control Commission by our division in August, 1959. No action was taken on these recommendations."

But now, early in 1964, modern science enabled the government teams to isolate the killing substance. Dr. Donald Mount, the Ohio State graduate who had toured the lower Mississippi observing the effects of the kill and picking up dead fish, is described as one of the heroes of the investigation. Working on a new method of "fish autopsy," he was able to uncover vital evidence in the fatty tissues of his specimens. Other PHS scientists in Cincinnati, using a gas chromatograph equipped with an electron capture detector, were able to assay extracts and blood from 100 poisoned fish between December 20, 1963, and April 1, 1964. Every sample contained Endrin. When infinitesimal quantities of Endrin, corresponding to the quantities discovered by sensitive instruments in extracts of the stricken Mississippi fish, were injected in healthy fish, the latter passed through identical symptoms and died. The next problem was to discover how Endrin got into the streams and rivers.

"Endrin is not found naturally in the Mississippi River," Murray Stein has said. "The fish don't go out to a supermarket or drug store or a package store and buy it. The fish must get Endrin from their total environment."

There was one clue. Although some scientists have said for a long time that insecticides were being washed off fields into streams and rivers, Louisiana officials believed that the big fish kill of November, 1963, was not caused by field drainage. For one thing, the fields are sprayed in the spring, not in the fall. For another, most of the Louisiana sugar-cane fields are cut off from the Mississippi and Atchafalaya by levees, thereby ruling out direct drainage into them; there were no big kills in the smaller streams near the cane fields. State Sanitary Engineer John E. Trygg summed up the state's position in his report of Endrin in the river: "Although the concentrations seem to increase in reaches of the Mississippi River in Louisiana there is really no drainage to the river from Louisiana soils and there are no Louisiana industries discharging insecticide wastes into the river."

Louisiana officials concluded that the great percentage of Endrin must be entering Louisiana from another state. It was on this basis (the pollution of interstate waters) that Louisiana had requested the assistance of the federal government.

Because the Mississippi River is the United States' major inland waterway, it is also its major sewer. Countless industries adhering to the old and once-valid adage that "dilution solves all pollution," have settled on its banks primarily to have access to a natural and therefore inexpensive sewer. Though there is not a major city along the Mississippi below Minneapolis and St. Paul which has an adequate sewage-treatment plant, no one had any qualms about dumping his most noxious wastes into the great but overburdened river. In 1964 St. Louis, Memphis, Vicksburg, Natchez, and New Orleans were still pouring their raw sewage into the Mississippi. New Orleans, as well as 100 other "jurisdictions" within Louisiana, with a total population of 1,094,000, take its *drinking* water from this cloudy sewer.

"Stream scientists working for Louisiana's Division of Water Pollution Control," Kenneth Biglane says, "soon become educated to the different types of water pollution that are found in their state from time to time. Wastes from sugar factories, sweet potato canneries, pulp and paper mills, oil field brines, naval stores plants, chemical plants, municipal sewages, and slaughterhouses all have two things in common. They can degrade water and they can kill aquatic life. Their point in time, their point source, and their physical and chemical alteration of the aquatic environment, however, offer clues to their dissimilarities."

Today there are about 10,000 industries in 18 states along the Mississippi and its tributaries. Of these, 118 plants manufacture pesticides. Before the development of new techniques and instruments by PHS scientists in the last few years, it would have been nearly impossible to trace a specific poison to a specific source. Now PHS investigators set out to find the source of Endrin in the Mississippi. After talking to agricultural people who had handled Endrin, and to scientists who had observed earlier fish kills in which Endrin had been suspected, they discovered an interesting fact. The effect on fish trapped in Endrin-infested streams near sugar-cane areas immediately after spraying and heavy rains had been "acute": the fish had died immediately. The effect on most of the fish during the recent kill had been "chronic": the fish had died more slowly or, in many cases, had gone through convulsions, righted themselves, and survived. Perhaps, the scientists thought, the more recent infestation of the river was of a different nature. Perhaps the poison had been diluted, after a long passage downstream, by mingling with the enormous flow of the lower Mississippi. At Memphis, Tennessee, some 500 miles up the river, there was a plant which manufactured Endrin. PHS investigators decided to take a close look at it.

The Memphis plant was owned by the Velsicol Chemical Corporation of Chicago. Endrin itself had been developed in Velsicol's Chicago laboratories. In discussing Endrin's widespread use in the South one agronomist put it this way: "I can say without the slightest fear of being contradicted that Endrin is the most effective insecticide ever used on the Louisiana cane crop." It is effective and popular, of course, simply because it is extremely deadly: it does the job.

In April, 1964, a team of Public Health Service investigators visited the Velsicol plant at Memphis. Leading the investigators was Dr. Alfred R. Grzenda, a pesticides expert from the PHS office in Atlanta. The story of what Grzenda and his associates found in Memphis might be enlightening to those who believe industrial pollution can be cleared up merely by "education and friendly persuasion."

Grzenda's job was to gather information on the manufacture of Endrin (including its by-products) and take samples of the water and mud in the vicinity of the plant. William Anthony, Velsicol's plant manager, was apparently reluctant to give Grzenda information about the various processes involved in Endrin's manufacture. "When we asked specific questions about starting and intermediate products," Grzenda said, "Mr. Anthony referred us to the patents for the manufacture of these compounds."

According to later newspaper reports, the PHS officials were barred from the plant, but Grzenda denied this. "We were treated well," Grzenda said, "but the Velsicol Company more or less selected the sites which we sampled. In other words, it was a guided tour through the plant."

Anthony showed Grzenda the plant's waste-treatment facility, a 50-ton lime bed used to produce a lime slurry which is added to a 40,000-gallon agitated retention tank. Anthony said that virtually no waste solids entered the tank. "However," Grzenda said, "sludge was dredged from the sides and middle of the tank and it contained a black, sticky, smelly residue. In the course of sampling, I splashed sludge on my face. It was extremely irritating to my eyes and skin."

Grzenda learned that the treatment plant was comparatively new, having been installed less than a year before (Velsicol had been manufacturing Endrin there since 1954, and discharging its wastes into nearby Cypress Creek until 1963). Asking for information on sewers in the area, Grzenda received contradictory reports from Velsicol people and city engineers about which pipes carried off trade wastes and which were city sanitary sewers. An attempt to enter one sewer was thwarted by noxious fumes; another attempt came to nothing when Grzenda discovered that smoke bombs planted in them by city officials ("to check for leaks") had made the sewers impenetrable. In all of them, including the "sanitary" sewers, Grzenda and his assistants detected strong chemical odors similar to that in the Velsicol treatment facility.

But the investigators did not have to rely on odors alone. Unexpectedly, they came across a by-pass line discharging Velsicol's trade wastes into a lake which flowed into Wolf River, a small tributary of the Mississippi. They also discovered that wastes, solid or semi-liquid, were being hauled daily from the Velsicol plant to a place called the Hollywood Dump. These wastes were carted there in caustic drums and in large containers known commercially as "Dempster Dumpsters." Shortly before the PHS men arrived, one of the workmen carting the drums to the dump had become ill when he was splashed with the wastes.

"This material is buried, or left exposed on a portion of the dump located on the flood plain of the Wolf River," Grzenda said. "All of the sites are subject to flooding. Mr. Anthony denied that any solids from the Endrin plant were being hauled to the Hollywood Dump, but we noted drums labeled 'isodrin scraps' at the dump on April 15. Isodrin is one of the compounds used in the manufacture of Endrin. However, Mr. Anthony said that such material is not normally taken to the Hollywood Dump for disposal."

Grzenda and his associates took samples of water and mud from the various sewers and dumps in the vicinity of the Velsicol plant, as well as from such small waterways as Wolf River and Cypress Creek (which he described as a "biological desert"). He found enormous quantities of Endrin and Dieldrin present. Calculating the amounts of Endrin which flowed from this area through streams and city sewers into the Mississippi, he was able to determine that about 7.2 pounds would enter the main stream in a single day, or 2,000 pounds a year. Although Endrin is marketed in a formula amounting to two percent, Grzenda made his calculations against "technical," or 100 percent, Endrin, which is 50 times stronger than the substance used on crops.

"I played with some figures that Dr. Mount gave me relative to blood volume and toxic concentration in the blood of catfish," Grzenda said, "and just playing around with a pencil, I figured if only one-*thousandth* of this amount of Endrin—that is if only two pounds a year found its way into the blood of fish, it would have the potential of killing 45 million one-pound catfish."

Grzenda, having completed his investigation, concluded that the concentration of Endrin in the water and underlying mud around Memphis exceeded by far all previous reports of concentrations of chlorinated hydrocarbon insecticides. Most of the contaminated areas were noted at sites "known to be downstream from points used or previously used by the Velsicol Chemical Corporation for waste discharge or disposal."

Soon afterward the Public Health Service made known its findings to the public. The presence of Endrin in the Mississippi had already been disclosed. This report was specific about its sources.

"Endrin discharged in the Memphis, Tennessee, area," the PHS report read; "other sources of Endrin not yet identified, and possibly other pesticides and discharges of sewage and industrial wastes of many kinds, pollute the waters of the lower Mississippi and Atchafalaya Rivers, and thereby, endanger health and welfare of persons in a State or States other than those in which such discharges originate. Such discharges are subject to abatement under the provisions of the Federal Water Pollution Control Act."

The Mississippi River, draining a vast area in the middle of our country, carries with it over 400 million tons of sediment a year. Reaching the sea it drops most of this sediment. Gradually the sediment, or silt, forms a delta. The 12,000-square-mile delta at the mouth of the Mississippi today has been built up over the centuries and is advancing into the Gulf of Mexico at the rate of 300 feet a year. Much of the south-central part of the United States is composed of land carried from the north and laid down by the Mississippi and its tributaries, so that the great river now flows through the land it helped to create. The river, then, is naturally discolored, but this has not always been a menace to the life it supports. On his voyage down the Mississippi, Huck Finn overheard a raftsman tell his colleagues that there was "nutritiousness in the mud, and a man that drunk the Mississippi water could grow corn in his stomach if he wanted to."

Whether there is nutritiousness in the river today is open to question, but there are a great many other things in it. The Mississippi, like every other major river in the country, has been blighted for years by raw sewage. The absence of treatment facilities in most of the cities and nearly all of the river towns is a disgrace with which the people have come to live. Raw sewage is the country's leading killer of fish. It also causes various serious diseases among human beings: hepatitis, poliomyelitis, cholera. The situation, until the recent mild spasm of treatment-plant construction, had grown steadily worse on the Mississippi. The process goes on from year to year, each town taking the river water, perhaps treating it and perhaps not, then using it and turning it back to the river, which carries it just a little bit more contaminated along to the next fellow downstream. The introduction of insecticides to the river in recent years has added another, and potentially more deadly, dimension to the problem. There is a morbid little joke currently circulating among pollution experts on the Mississippi. "Pesticides can't do anything to our fish this year," it goes. "Our sewage has killed them all."

It was not a laughing matter in New Orleans when the Public Health Service released its report on Endrin in the river. Traces of Endrin were found in the drinking water of that big river city, as well as in the water of Vicksburg. (Purification systems for drinking water cannot remove Endrin.) New Orleans citizens who could afford it took in supplies of bottled water. Only a few hours after the PHS announcment, one New Orleans bottled-water dealer announced that he was placing customers on a five-month waiting list for his product. The uproar continued throughout April. James M. Quigley, Assistant Secretary of Health, Education and Welfare, said that after looking over the evidence he would not care to eat a Gulf shrimp cocktail. Dr. James M. Hundley, Assistant Surgeon General, on the other hand, said that he didn't feel there was any present danger involved in drinking Mississippi water or eating Gulf shrimp, but he added that he did not think it healthy to try to subsist exclusively on a diet of Mississippi catfish. The Food and Drug Administration discovered an interstate shipment of canned oysters containing traces of Endrin, but did not seize it on the grounds that it was not considered "an imminent danger to health." The government decided to survey the entire season's harvest before deciding what to do with Endrin-contaminated samples. Secretary of the Interior Stewart Udall called for an immediate ban from agricultural use on long-lasting pesticides (and ordered such a ban throughout the 550 million acres of public land administered by his department). The *New York Times*, in an editorial, supported Udall's position.

"These pesticides do not break down in nature," the *Times* editorial said. "They retain their potency long after their initial use. Once put into the environment by farmers and others, these chemicals tend ultimately to enter the food chain of living creatures. There is evidence suggesting that the degree of their concentration in organisms increases as contaminated fish, for example, are eaten by birds. Since human beings complete such food chains, there is some ground for suspicion that such increasing concentration can endanger human health.

"Here is a situation," the *Times* continued, "in which the case is strong for imposition of at least some controls quickly while research goes on to accumulate more information and also to find substitute pesticides which do not pose so formidable a threat. Certainly the nation's overcapacity for food production is so great that any potential diminution of crops resulting from such pesticide restrictions would be a lesser risk than that arising from the haphazard use of the chemicals themselves."

Velsicol, under fire, shot back. Bernard Lorant, the company's vice-president in charge of research, issued strong denials. In a statement to the press, he said that Endrin had nothing to do with the Mississippi fish kill, that the symptoms of the dying fish were not those of Endrin poisoning, and that Velsicol's tests proved that the fish had died of dropsy. He went on to question the Public Health Service's accuracy in analyzing fish samples, claiming that, since Dieldrin is not a by-product of Endrin's manufacture, Dieldrin traces could not have been found at Memphis.

The row shifted from the pages of the daily newspapers to the hearing rooms of various committees investigating the fish kill. Hearings were called by both Senator Abraham Ribicoff's Sub-committee on Government Operations and the United States Department of Agriculture. Velsicol continued to question almost every aspect of the Public Health Service's case. Lorant pointed out that the amounts of Endrin in question often amounted to only a few parts per billion. He said that, of the 5,175,000 fish reported killed, five million were menhaden ("which is 96.6 percent," Lorant said, "or in the vernacular of Public Health, 966 million parts per billion"); these menhaden died in the mouth of the river, he went on, and were not collected or sampled by PHS; as there were only 175,000 others, mostly catfish, Endrin could be blamed at most in the death of 3.4 percent. He asked how Endrin could drift 500 miles down the river without causing a fish kill on the way, asked why no fish kills had been reported during 1961 and 1962, although Endrin was being manufactured and applied just as regularly, and said that it would have been impossible for Endrin to contribute to pollution in the lower Mississippi by running off the Louisiana sugar-cane fields, because the fields were cut off from the river by levees.

Velsicol was supported in its arguments by the Shell Chemical Corporation, the only other manufacturer of Endrin. According to a Shell executive, Endrin is readily absorbed into mud. "It is reassuring to realize," this official said, "that the large silt load carried by the Mississippi River acts as its own 'clean-up' agent for chlorinated or other highly absorbable pesticides."

But the Public Health Service stood its ground. One witness pointed out that the so-called infinitesimal quantities of Endrin found in certain areas (measured in parts per billion) are not quite so negligible when one considers the tremendous flow of the Mississippi. Others remarked that there was no major fish kill near Memphis because the fish in Endrin-infested waters had probably been killed off long ago ("Sometimes a sure sign of pollution abatement is a fish kill," one government investigator said. "At least it proves that there are still fish in the vicinity"). Endrin's report that the fish kill could be attributed to dropsy, and that there had been no reports of a kill in 1961 and 1962, were countered with the reminder that older methods of analysis had not been able to detect Endrin, and that earlier fish kills were not reported since they were considered due to natural causes.

In reply to Velsicol's statement that Endrin could not have reached the Mississippi from Louisiana cane fields, one PHS scientist suggested that the Endrin might come from factories where the cane was washed and treated; some of these plants border the Mississippi.

Witness after witness reflected the alarm that had invaded many sections of our government. It was noted that British authorities had already placed restrictions on three pesticides which have poisoned fish and shrimp in the United States. Dr. Clarence Cottam, a noted biologist, warned, "We're going to find human beings dying of this thing, unless we act with intelligence now." Assistant Secretary of HEW James Quigley testified, "The presence of any of these materials in any food or liquid consumed by human beings is a cause for concern, even though the levels may be far below those which might be considered an imminent health hazard."

When asked what amount of Endrin might be lethal to human beings, Assistant Surgeon General Hundley confessed to Senator Ribicoff's subcommittee that he did not know. "I don't know that an answer is available," he said.

After hearing the evidence that Endrin was present in shipments of shrimps and oysters, Senator Ribicoff commented, "There must be an awful lot of fish going into interstate commerce that should be condemned and removed from the market."

But the reply of George P. Larrick, Commissioner of the Food and Drug Administration, was guarded. "That remains to be seen," he said. "I don't want to just destroy the market for those products . . . I would prefer to wait until we have run a lot of samples. I hope we won't find massive contamination, and I don't think we will."

Ribicoff, after hearing all the testimony, came to some strong conclusions. He told the Velsicol people that they had disposed of their chemical wastes "in a primitive and dangerous manner ... The record is overwhelmingly against your position."

Officials in the Department of Health, Education and Welfare came to similar conclusions. Until that time they had not wanted to restrict the use of dangerous pesticides, but the findings in the Mississippi River had changed their minds. "We cannot proceed in the same way in the future as in the past," Assistant Secretary Quigley testified before Ribicoff's subcommittee. He said that HEW now believed that as a "matter of prudence every possible effort should be made to control the use of the persistent pesticides in our environment."

The Department of Agriculture had also found disturbing evidence of misuse or careless handling of insecticides. Its representatives had accompanied PHS investigators to the Velsicol plant at Memphis. A USDA report confirmed that Velsicol's disposal of its wastes "is contributing substantially to the contamination of the river. It is our understanding, however, that the waste disposal system employed by this plant is in compliance with local sanitation codes."

A dispatch printed in the *New York Times* on January 17, 1965, reviewed some of the other curious circumstances in Velsicol's recent history at Memphis. "On June 3, 1963," the *Times* story said, "the Memphis Health Department reported complaints from 20 persons living near Cypress Creek, an open stream flowing through the north side of the city. Nausea, vomiting and watering eyes were the symptoms produced by gas rising from the stream. An official of the Velsicol Chemical Corporation denied responsibility. 'Endrin could not have caused the symptoms,' said Wilson Keyes, director of manufacturing.

"On June 7, 1963, 26 workmen in plants near Velsicol were taken to five hospitals after becoming ill from chlorine gas fumes. Within a year lawsuits totaling over $5 million had been filed by more than 40 persons claiming injury.

"Velsicol reacted in the first weeks of the trouble with a dinner for 150 political, civic and business leaders. 'It came as quite a shock to us to discover that there was some question about whether we were welcome in the City of Memphis,' said John Kirk, executive vice-president, down from Chicago for the event. Then Mayor Henry Loeb responded that 'this plant is very much wanted by Memphis.' His successor, William B. Ingram, Jr., took much the same position a year later."

Elsewhere along the Mississippi, Department of Agriculture investigations into pesticide plants proved to be "quite revealing." The USDA report said that "conditions were observed which appear to constitute a definite hazard. For example, it was found that a cooperage company removed material from used drums by heating and then washing them with caustic soda. Periodically, this solution is flushed into the city sanitary system. It was also found that several plants dispose of wastes at city dumps and that as a rule these dumps are located on the river side of the levee."

But the USDA, with the farmer's welfare uppermost in its reasoning, refused to take restrictive action. It confirmed the Public Health Service report that large quantities of Endrin and Dieldrin had been found in the Memphis area, but said that the river contamination had nothing to do with the agricultural use of these insecticides. It called, as usual, for further study. A Public Health Service scientist noted that the USDA, which is charged with the growth and welfare of American farming, continues to move into health research as well. "It's like getting a jewel thief to guard the jewels," he said.

The full-dress parade of all the combatants was reserved for the conference which was held on May 5-6, 1964, at New Orleans. It was called by Anthony J. Celebrezze, Secretary of Health, Education and Welfare, as a direct result of Louisiana's plea for help and the subsequent investigation by various government agencies. Present at the conference were members of these agencies, representatives of Arkansas, Tennessee, Mississippi, and Louisiana, and a number of other interested parties. Murray Stein, as usual, served as chairman. If the conference was not a complete success, at least it dramatized the obstacles to enforcing an effective program against water pollution in this country.

The Public Health Service, occasionally calling on representatives of the Department of the Interior and the Department of Agriculture, presented its lengthy case and its conclusions. Its investigators told the conferees what they had found in Memphis, and its scientists painstakingly reviewed their methods of analysis, and the results of their tests. Velsicol made its defense, and was rebutted by PHS.

Then the states and other industries took their turn before the conferees. Going over the transcript of the conference, one is impressed finally by the doggedness with which each speaker strove not so much for a final solution of this complex problem, but with defending his own position and interests. Those whose concern was with fish or other wildlife pleaded for immediate action. A representative of the Tennessee Fish and Game Commission, after describing the carnage inflicted by insecticides on the state's fish and birds, recommended that "the manufacture, distribution, and use of Endrin, Aldrin, Dieldrin, and other related chlorinated hydrocarbon pesticides be banned until the full effects of these poisons can be determined and evaluated." On the other hand, a representative of the sugar-cane growers asked the PHS to take no action, or, in a euphemistic phrase, to "undertake an extensive survey and study program to determine all the facts in this matter before a decision is made that could possibly do an injustice to agriculture in general and weaken our nation."

The tail end of that phrase struck a chord that seemed to be popular at the time with those under fire from the Public Health Service. "The great fight in the world today is between Godless Communism on the one hand and Christian Democracy on the other," Parke C. Brinkley, president of the National Agricultural Chemicals Association had said earlier that spring. "Two of the biggest battles in this war are the battle against starvation and the battle against disease. No two things make people more ripe for Communism. The most effective tool in the hands of the farmer and in the hands of the public health official as they fight these battles is pesticides."

It was a chord struck repeatedly during the New Orleans conference, as backward, Godless, pesticideless Russia was compared to a flourishing America. Another popular gambit used by Endrin's defenders was to try to make the Public Health Service ashamed of itself for doing its duty. "Until there is evidence to the contrary," a Shell Chemical official said, "it would be irresponsible for any reputable person to frighten people into believing that eating catfish or similar game fish is dangerous."

Velsicol's Bernard Lorant compared PHS's warnings to shouting "fire" in a crowded theater, then sounded aggrieved that the investigation had heaped extra work on the scientists. "None may allow unfounded, unthinking utterances that trigger a massive response of waste," he said. "In the present climate, should anyone recklessly conjecture as to hazard, hundreds of scientists initiate programs to prove or to disprove the remark. We do not always need more scientists; we frequently need less conjecture."

Representatives of the various states taking part in the conference apparently were more concerned with the ogre of "Big Government" than they were with the threat to their waterways. Though the conference was called because one sovereign state had been powerless to prevent pollution from pouring across its border, the representative of another state felt called upon to include among his recommendations to the conference a request that the federal government begin to know its place.

"It is not possible for the government in Washington to be all things to everybody," said E. H. Holeman, of the Tennessee Department of Agriculture, "and especially in the area of protecting wildlife, in preventing stream pollution, and in protecting the health and welfare of the consumer.

"The federal government will have to recognize this fact, and it is urged that they closely coordinate their consumer protection program with the state officials in order that we will have much better and more strict and a more harmonious consumer protection program."

This aggressive attitude toward Washington always carries with it a counteraction, much like the double standards of those foreign governments which encourage the stoning of our embassies and libraries one day, and ask for handouts the next. As in the present case, the most ardent States' Rightists are always quick to call for Washington's help when trouble crops us. A few minutes after Holeman had put forth this "recommendation," conference chairman Murray Stein pointed to another recommendation in which Holeman had asked for a federal standard on the limits of pesticides to be permitted in marketable foods.

"Aren't you worried about a federal standard usurping states' rights?" Stein asked him.

"Oh, no," Holeman said.

"Even though this might affect water pollution control and you would have a standard coming from Washington?"

"That wouldn't bother us. We have the same law in Tennessee, and so do 36 other states, as the Federal Food and Drug Act."

"It is very interesting to get the view of Tennessee on that," Stein commented.

At the conclusion of the conference, as Stein was trying to get the conferees to agree on a summary, the proceedings nearly bogged down in a spasm of local pride. When Stein attempted to insert a phrase saying that a certain condition "may require further study," a couple of Louisiana representatives put their backs up.

"Obviously, it is true," one of them told Stein, "but I don't like to see it there. It is an invitation for you to come back, you know. We don't think we need you on this."

"I understand," Stein said.

"We have tried not to worry you for about three years, as a matter of fact," another said.

The word *must* set off another furor. When Stein suggested the inclusion of the phrase, "known sources of Endrin must be brought under control," S. Leary Jones of the Tennessee Stream Pollution Control Board spoke up.

"I object to these *musts* coming from a federal agency," Jones said. "Make it *should* or *ought to be."*

"How about *are to be?"* Stein asked.

"We are going to clean up the stuff," Jones said, "but they can't tell us what we *must* do."

"The thing is this," Louisiana's Robert LaFleur, who had made the original telephone call to Washington, interrupted. "If we are going to suffer from this next fall, I am going to be boxed into a corner, and I don't want that."

"This won't happen, I guarantee," Jones said. "But a *must* in there is just one of these words that I won't agree to."

"I sure want it to be cleaned up," LaFleur said.

"All I can tell you, Bob, is that if we say *should* it has no force and effect," chairman Stein reminded him.

"It doesn't have any force or effect either way, until you go into a hearing and into federal court," Jones said. "You know that."

"Most of these cases have been solved by conference," Stein reminded him.

"This isn't a case. This is a conference."

"When I speak of case I am using a generic term," Stein said.

"All right," Jones said. "Put *should* there."

"If that man will promise me he is going to clean it up," LaFleur nodded, "I will buy his *should*."

And so the conference sputtered to a close. Endrin, originating at Memphis, was concluded to have been a contributing factor to the previous fall's kill of Mississippi catfish, and all parties were urged to see that it did not happen again. Yet, though the government's finest scientists, equipped with the most modern machines and techniques, had come to a definite conclusion about the source and agent of the fish kill, the uproar continued. Velsicol went on maintaining its (and Endrin's) complete innocence. *Chemical Week*, a trade publication, called the conference "a kangaroo court." And, on the floor of the United States Senate, Republican leader Everett McKinley Dirksen rose to attack the United States Public Health Service, claiming it had made "wild accusations" and had "unjustly crucified" Velsicol before it had all the facts. Dirksen's tirade surprised nobody. The Illinois Senator has been for a long time the champion in Congress of the nation's drug and chemical companies, as anyone can testify who recalls his bitter opposition to the Kefauver drug bill a few years earlier.

There were expressions of dismay in the press and among public health and conservation groups when Velsicol persisted in its unregenerate stance. On the other hand Murray Stein, a man of few illusions, speaks of the case today with some satisfaction. "A lot of the criticism from the chemical people was directed at me personally," Stein says with a delighted grin. "They made a big man out of me and I got some offers for important jobs. More important, though, in many aspects this was the easiest case we've ever had. It was open and shut, and our tests show that the river around there is a lot cleaner since we went after those people."

In 1965 a few fish died in the upper cane brakes after Endrin had run off the fields in the spring, but there was no slaughter on the scale of previous years. At Memphis, PHS investigators looked into a city sewer, however, and found it caked with depostis of Endrin sludge nearly three feet thick. It was estimated that 8,000 pounds of Endrin were embedded along a 3,400-foot stretch of sewer. City officials hastily closed it off and built a bypass sewer.

"The stuff is still in the sewer," Stein says. "And it will probably stay there for a while. Who would go in there after it?"

Malvina Reynolds (1900-   ) is a song writer whose lyrics are recognizable for their humorous and ironic attacks upon social ills and injustices. She has a Ph.D. in English from the University of California at Berkeley and is active in many conservation groups. This essay reflects both her love for San Francisco Bay and her indignation at its threatened destruction. The gradual filling of the bay began in 1860, and its area has been reduced from an original seven hundred square miles to about four hundred square miles. The argument that population growth is the root cause of environmental destruction is made clear here: Real estate developers and the Army Corps of Engineers, in response to the growing demands for land from a burgeoning population, fill the bay's marshes, mud flats, and open water, making room for more people who then create more highways, waste, sewage, etc. The marshes, mud flats, and open water do have great intrinsic value in an ecological sense, but this is lost sight of in the scramble for profits and so-called progress. Only recently has the public begun to recognize the bay as a natural resource rather than as real estate.

A SONG OF SAN FRANCISCO BAY

by Malvina Reynolds

The highway is laid as smooth as glass
For miles and miles and the cars can pass.
But the ant and the bee and the bush and the tree
Whose home it was are now exiles.
And the cars rush by for miles and miles,
To find a place where they can see
A plant, a bush, and a blade of grass,
And a ladybug, and a bee.[1]

Of the 270-odd miles of shoreline of San Francisco's great bay, some—not much, but some—is still in its natural state: grassy or lightly wooded hills leading down to a rocky border; marsh and tidelands green with reeds and salt grass, still alive with sea and shore birds and the infinite smaller creatures who populate such places. And if you wanted to see something like this, you might take a thirty-mile drive along the bay's northern shore—say on Highway 37—between Vallejo and San Rafael, with no towns between. If you'd been there before a few years ago, you would remember a long stretch of secondary road a few

[1]From "The Highway," words and music by Malvina Reynolds. Copyright © 1965, by Schroder Music Co., ASCAP.

feet above tideland level, with open marsh or farmland on both sides and the bay itself stretching to the south.

But now on that route you would find no bay, no marshes, not even farms for miles beyond Vallejo. Just the usual decorations of the modern highway—auto junkyards, hamburger stands, motels, small factories, billboards, gas stations—and the bay not even in sight. So you might take one of the unnamed roads that heads south to where the bay should be, and you could find yourself in the enormous grounds of the Kaiser Steel plant. Driving into the grounds you'd go over a bumpy road through an expanse of raw ocher earth, with a lonely bunch of reeds in a small pool at a culvert showing what had been here before the steel mill took over. As far as the eye can see, lines of flatcars on the spur track carry monster steel pipe probably destined to provide drainage for the new highways that are being laid everywhere you look.

This isn't what you were looking for, so you bump back on the branch road to 37, into the roaring raceway of cars, heading westward again. It is ten miles or more before you finally reach open fields and then the marshes, with reeds and salt grass and, for the first time this trip, the cool wind that blows from the bay and across the flats.

After a while you want to stop, but it isn't easy to turn off, with the heavy traffic averaging 50 miles an hour, and some people might wonder why anyone would want to stop here anyway. No garage, no hot dogs, no bar—the cars all tearing by, headed some place. But it happens there are white herons feeding not far from the road, and you hope to get them with your camera. They have become accustomed to the heavy trucks roaring by and the endless line of cars. But you have to come up easy with the camera—they are not accustomed to people on foot.

Seventy miles of wind and spray,
Seventy miles of water,
Seventy miles of open bay—
It's a garbage dump.[2]

I wrote the song "Seventy Miles" in March of 1965. For me, it marks the first time I became sharply aware of what was happening to San Francisco Bay—my bay, everybody's bay.

---

[2]From "Seventy Miles," words by Malvina Reynolds, music by Peter Seeger. Copyright © 1965 by Abigail Music Co., (BMI).

In a way, San Francisco, port to the coast trade and the Orient, takeoff point of the gold rush of the 1850's, commercial center of the mineral and agricultural wealth of the state, is everybody's city. Will Irwin's *The City That Was*, a little "requiem of old San Francisco," written in 1906, three days after the great earthquake and fire, opens with a quote from one Willie Britt, "I'd rather be a busted lamp post on Battery Street, San Francisco, than the Waldorf-Astoria." And Irwin himself describes the city of those days as "The gayest, lightest hearted, most pleasure loving city of the western continent." Fifty years later, Joseph Henry Jackson, book reviewer of the *San Francisco Chronicle*, opens his booklet tribute *My San Francisco* saying, "San Francisco is a great and greatly loved city."

Songs are written about San Francisco in every generation; visitors who come here go back home to pack their belongings and get out here to live. Whether they came with the covered wagons, with the railroads, with the gold-rush pioneers; as servicemen in many wars, as workers in war industries, or just as summer tourists running away from the muggy heat of other parts of the country—they settled here at last. Areas around the metropolises of the West Coast are among the fastest growing communities in the world.

And even if they don't get here to live, if they have only visited or if San Francisco is only a name to them, they feel a sense of identity with this city; it is part of the romantic tradition of America and the West. But I was born here. Grandparents on my mother's side were pioneers who left Lithuania in the 70's and, rejecting the East Coast, came directly to San Francisco. My father came alone from Budapest when he was fourteen. I was born south of Market Street in San Francisco—now an area of small industries and wholesale houses, but then a respectable lower-class section of wooden railroad flats, a few of them still standing among the warehouses and plants. I have lived in this city or near it most of my life. As the saying goes in this town, "Why should I go anywhere? I'm here already."

The bay and the hills around it are part of my natural environment, as they are of most of the three million who now inhabit the bay counties. Until about 1965, I had taken this great inland sea for granted. But all the time, it was being edged in on.

One of the maps in the office of the San Francisco Bay Conservation and Development Commission is a colored layout of the bay area showing the landfills that have been made since 1860. By 1900, when I came on the scene, the three coves that had sheltered sailing ships on the eastern side of San Francisco peninsula had been largely filled and built over. Later, while earning my degrees—and learning how to have fun—at the university in Berkeley, agents for farm land were filling about twenty square miles of mud flats and marsh in the northeastern arm of the bay; and under my eyes, as I crossed back and forth on the new transbay bridge, a whole new island had been laid in the shallows next to Yerba Buena Island, to accommodate first a world's fair, and then a Navy base. Some square miles on the eastern shore, south of Oakland, had also been filled.

In 1940, my husband Bud, a construction worker, was up to his ankles in mud, working on the shipyards that would be part of a continuing extension of the town of Richmond, jutting out from the bay's east shore. In 1942 and 1943 he worked on the Oakland Naval Base, which, with the adjacent Alameda Naval Air Station, thrust fill and docks for a mile or more into the bay alongside the East Bay bridge approach. At the same time the San Francisco airport, built on fill at the south end of the bay, was granted an extension of submerged lands. Ten years later, 7,000 acres of tideland would be granted to Oakland for its airfield. Two beautiful small bays on the northwest shore, in Marin County, had been converted from jewels reflecting Mount Tamalpais and the wooded foothills into flat, dirt-filled areas with a few buildings for decoration.

Free fill supplied by the Army Engineers from the dredging of the bay bottom was part of the material used to make these continual fills. More came from hills that were being leveled for factories and housing sites. The most offensive was garbage. At this writing, the city of San Francisco still dumps its tremendous daily load of garbage into the southern end of the bay, in spite of a valiant fight by people of the city of Brisbane, on the face of San Bruno Mountain and overlooking the flats, against the offense in their front yard. There are more than thirty refuse disposal sites in and at the edge of San Francisco Bay, from Martinez and Pittsburg on the northeastern arm to Palo Alto and Alviso at the south. And the discharge points for sewage and other fluid wastes, domestic and industrial, are about three times as many.

Because the mountain formations of the West Coast run generally north and south, there are very few good harbors on the Pacific. San Francisco Bay is the largest and finest south of Puget Sound, a water area at mean high tide of more than 435 square miles, almost completely protected from the sea except for the strait on the west called the Golden Gate. The bay itself lies in what was a long valley, extending north and south. Into its northern arm empty two big rivers of the Central Valley—the Sacramento, flowing south, and the San Joaquin, flowing north.

More than 400 square miles of open water. But when California was admitted to the union, before any "sanitary landfill" had been dumped into the bay, its area was more like 700 square miles. If you tell the average San Franciscan that the bay is one-third smaller than it was when discovered, he will be startled. And if you tell him that a good deal of the rest of the bay—and I am talking of water-filled bay as far out as the deep-water boating channels—is city, county, state, or privately owned real estate, marked into lots and changing hands at ever rising prices, he will question your sanity. But it's true. Mel Scott, Research City Planner commissioned in 1963 by the Institute of Governmental Studies of the University of California at Berkeley to do a study on the future of San Francisco Bay, tells of the ownership of these submerged tide and water lots in the bay: "Almost the entire waterfront of Sausalito has been consolidated by a single syndicate. Just seven owners now control the submerged properties in Corte Madera Bay, among them the Marin Title Guarantee Company, the Utah Construction and Mining Company, the Wells Fargo Bank and the City Title Company. . . . On the eastern side of the bay the Santa Fe railroad has title to almost all the privately held tide and submerged lands from Richmond to Oakland, and in Richmond alone it owns 1,156.13 acres. The Standard Oil Company possesses more than a thousand acres of offshore properties in Richmond, and claims title to another 640 acres." And so on and on. I sat on the deck of my friends' pleasure boat in the Sausalito Yacht Harbor on a pleasant Sunday afternoon, and they pointed to a section of the bay, far from shore, where yachts and trimarans were sailing by. "A lot out there," they said, "just changed hands at a price of $600,000."

"For eleven decades," says Scott, "the state government has tended to regard the bay as property rather than as a great natural resource to be safeguarded." While the state still owns the greater part of the northern arms of the bay—Suisun and San Pablo bays—it is in possession of less than half of the main body of the bay.

This process of local and private occupation of underwater real estate and tidelands began even before California became a state, with an edict issued in 1847 by Brigadier General Stephen W. Kearney, military governor of the territory, for the sale of submerged lands in Yerba Buena Cove at public auction—"proceeds for the benefit of the town." The cove exists no longer, but is filled from Montgomery Street, San Francisco's Wall Street, to the Embarcadero, the wide avenue along the bayside docks. It now supports a sizable part of the downtown city.

They say the filling process began, perforce, when ships of the 49 Argonauts were abandoned in the cove, a whole fleet that had come around the Horn and anchored there—the first no-deposit, no-return throw-aways in the state's history. Some of them sank and are part of the underlay of tremendous office buildings; others were used as temporary dwellings and business places in the exploding town of the mid-century, and the area around them was filled for foot and carriage passage. The ship *Apollo* became a saloon; the brig *Euphemia* served as a prison ship.

At any rate, the process of private ownership of the bay was made official by the Army officer who is generally forgotten even by those who use busy downtown Kearney Street, which is named for him. And the process has continued under confused authority, until we have the situation described by Mr. Scott, with much of the bay now in title deeds. The agencies trying to hold the line against filling of the bay have to battle these holdings, foot by foot.

At first thought, to a tenderfoot like me, much of the bay would seem to be of questionable value in its native state. A good deal of it is quite shallow even at high tide, and the U.S. Army Corps of Engineers, with the engineer's practical eye, describes 248 square miles of tidelands as "susceptible of reclamation."

You will excuse a poet's digression here. I am sensitive to words like *reclamation* and *development* because they are not, as they would seem to be, impersonal and descriptive. They are highly charged with propaganda. My dictionary says of "reclaim": "to bring back to a useful, good condition," and two common uses of the word are given in the quotations that follow: "The farmer reclaimed the swamp by draining it." "Henrietta reclaimed him from a life of vice."

We are to believe that the dumping of waste and garbage into the edges of San Francisco Bay, the covering of it with the bedrock of ravaged hills, is a reclamation.

The building of housing and businesses would at first thought fulfill the idea of "development," which Webster's Twentieth Century Dictionary describes as "the causing to become gradually fuller, larger, better."

But the development of the bay region has one immediate, adverse effect. In a crowded part of the world, it puts people in residences and in factories where there were no people before. All the harmful results of overpopulation are immediate—increased strains on waste disposal facilities, increased appearance of pollutants in air and water, increased demands for highways and parking facilities, increased psychological damage from crowding.

But the shallows and mud flats in their natural use have values of their own. It does not take an expert to know that open space is a great value in itself. When it carries bay and ocean breezes, sweetened with oxygen and natural moisture, it is even more precious. And the mud flats and marshes are in themselves producers of oxygen. Mud algae, exposed to sun and water, produce oxygen for air and water. So do the marsh plants, the tules and the marsh grasses. The latter are very rich in food value for the sea and bay creatures that feed on them as the plants die away. Cord grass, one of the marsh plants, has seven times the food value of an equivalent acreage of wheat, and in the cycle of life that goes on in the waters of the bay and the ocean, these food values insure a rich crop of fish and other sea food for the multimillion-dollar sport and industrial fisheries.

I'm sorry to find myself talking in money terms. But every agency that fights for preservation of our bays, rivers, and forests finds itself having to appeal in such terms to the governing bodies—city, state, and national—that have these decisions in their hands. Otherwise, they can be accused of "catering to the bird watchers."

The bird watchers—there's a laugh.
Their idea of having fun,
Sitting on a hillside under the sky,
Sensing the trees and feeling the sun,
Watching the birds who nest and fly.
Watching the castle clouds go by.
Watching the flowers, watching the bee.
When they could be sitting at home with a beer,
Watching T.V.

Besides being the home of many local characters—sea gulls, pelicans, herons, cormorants, and loons—San Francisco Bay is the stopping station for hundreds of thousands of birds on the Pacific flyway, a run that extends from South America to the Arctic Circle. The marshes and waters of the bay, where the water is less than eighteen feet deep at low tide, are the principal feeding, resting, and wintering grounds for seventy-five different species of water birds that visit and live here. Mussels, clams, snails, worms, and insects are the main of their diet; the deep waters of the bay are their playground and their shelter during storms at sea. The bay has a great fish population, too, and many species that live in the sea come to these shallows to spawn. The young feed on the marsh and mud flat life, and vice versa.

There are about fifty square miles of marshland remaining around the bay, sixty-five square miles of tidal flats, and seventy-eight square miles of land at the south end of the bay that are used for salt production—commercial, indeed, but still an open area and one that sustains certain water creatures adapted to the concentrations of salt in the drying pools.

But if, in your scenic drive, you missed the way to Highway 37 and came to the new bridge over the Napa River, which feeds into the bay from the north, you would find inroads into the open areas—new highway being laid over the marshland; raw yellow earth over a wide swath of what had been reeds and pools and small sea creatures. A poet sitting by the driver of the car might have scribbled in his notebook:

Goodbye to the reeds and the little fishes,
And the shouting boys who play under the bridges.
Raw earth dug from the doomed hillside
Pours over the marsh grass, child of the tide;
The living wind that blew from the bay
Now carries the fumes of the swarming highway.

And at scores of places on the bay shore, and notably at the south end, eight thousand tons of assorted garbage—trash and the stones and waste of demolished buildings—are daily dumped into the bay.

We are generally an affluent and wasteful people, and each one of us produces his four pounds of discards a day—not to mention sewage and the products of industrial waste that our consumption of stuff requires.

What to do with all this? I don't know. I only know that a civilization that can send men to the moon can certainly find other ways of handling this problem than by destroying a city's greatest natural resource for a dump. I read somewhere about a European city that consumes its waste under very high heat, and presses the resulting slag into building blocks. What a great idea! If this is so, then maybe it doesn't have to fell its forests for building materials, either. Apparently there are many kinds of engineers in the world.

The tone of my discourse, so far, would indicate that I favor pelicans, wild geese, striped bass, and sea snails over human beings. I would hate to be pushed too far on this question, but it is really not relevant, because people and pelicans alike live much better lives as one in the life-chain with marsh grass, ladybugs, forests, robins, wild blackberries, and shrimps. If more people were packed into the space already marked off in the bay as real estate, they'd all be worse off then if they had never come here in the first place. California is a tremendously large state, and there are millions of acres that could absorb any normal increase in population. That is, if we weren't all city-oriented, which is another question.

The great pulse of the ocean tides sweeps into San Francisco Bay twice a day, and ebbs out, to create a tremendous circulatory system, which cleanses the bay waters, aerates the waters and mud flats, and feeds the creatures who depend on this tidal change for their environment. Meanwhile, fresh water from the Sacramento and San Joaquin rivers flows into the north end of the bay, sometimes over the salt water beneath. In the central area of the bay the ocean tides can often be seen moving in over the underlying fresh water, and underneath are tremendous currents caused by these pulls and drifts, as well as by the contours of the bay's edges and bottom. The oxygen content of the water is affected by these movements, and on this content depends the ability of the waters to handle and convert the outflow from the cities, to prevent the waters from becoming lethal to wildlife and people. Every reduction in the area of the bay works against this function, and about now, the margin of safety is pretty small. Four and a half parts of oxygen per million of water are necessary for the survival of the fish and other creatures of the bay waters. It now stands at an average of something like seven to nine and a half parts per million—and the garbage trucks and dump trucks are running down the Bayshore Freeway all day, every day, while the Engineers offer free fill to all who want it.

According to the San Francisco Bay Conservation and Development Commission, if all the relatively shallow parts of the bay shown by the U.S. Army Corps of Engineers to be "susceptible of reclamation" were filled (and, as we have seen, about 25 per cent of the bay, including most of the shallow areas nearest the shore, is claimed by private owners), the bay would consist of only 187 square miles. In some places it would be little more than a river. The rest would be what Harold Gilliam, feature writer of the *San Francisco Chronicle*, calls, in an article in the Sunday magazine of March 7, 1965, a "fuming flat." The local provincial patriots among the conservationists fearfully describe the resulting place as "another Los Angeles." I am something of a neutral in this conflict between cities, but I think they may have something.

Anyway, what I think of the fill projects—and every county, every owner of submerged land in the area, has such projects in blueprints—opens the song, "Seventy Miles."

> *What's that stinky creek out there,*
> *Down behind the slum's back stair,*
> *Sludgy puddle, sad and gray?*
> *Why man, that's San Francisco Bay!*

Harold Gilliam, quoted above, is one of the great naturalists of the Pacific Coast, and a favorite columnist on San Francisco's only major morning paper. His work on San Francisco Bay, published in 1957 by Doubleday, is a learned and poetic tribute to what he calls "the incomparable harbor." Each section of the book opens with a lyric painting of a particular aspect of the bay, its hills, and its sky.

"Clouds hang low over the vast amphitheater of the bay and its shores like a pavilion roof supported by mountain pillars. The scene is a monochrome of grays of infinite shadings from silver to near-black. The morning air is still, and the bay is flat and glassy—a clean scroll ready to be written on by the winds, the fogs, the rain, and passing vessels.

"A freighter leaves a spreading V-shaped wake of waves and ripples sharply etched on a surface of gunmetal gray. A tug crosses the ship's wake and sets up a conflicting series of wave lines. . . ."

. . . . .

Too bad, too bad, that we have to talk of garbage, and the proposed tearing down of a mountain (the only one in the southern area of the bay, San Bruno Mountain) for fill to eradicate another swath of this great inland sea. It is too bad that we have to talk of sewage outlets, of "industrial parks" and freeways and housing developments projected for the precious wild spaces that still remain and even for the deepwater areas within the bay.

But I am not the only one by a long way who "came to" in the early sixties as to what was happening to my bay. I discovered, when I looked around, that many conservation organizations were actively concerned, that leading citizens of the area had formed a Save the Bay Committee, that the counties adjoining the bay had realized that some measure of joint planning was necessary to prevent the disaster that they were all beginning to foresee otherwise. There is now an Association of Bay Area Governments, a Bay-Delta Water Quality Control Program, a Citizens for Regional and Recreational Parks, and under the aegis of the state legislature, a Bay

Conservation and Development Commission. There is a moratorium on filling—not complete, as we have seen, but a hopeful start—while BCDC prepares and presents its recommendations. The concept of development is included, granted—but at least some of this development seems to be in the direction of rescuing the few miles of publicly owned bay shore for public parks, adding to these miles by purchase where it's possible.

The eastern arm of Richardson's Bay, one of the bays within the larger bay, was recently threatened with development as a private residential marina, but conservationists raised a quarter of a million dollars, bought the place, and then turned it over to the Audubon Society as a wildfowl sanctuary.

Though there are constant pressures for easy airfield extensions, building sites, highways, and dumps into this great open bay's shallower areas, there is hope that most of it can still be rescued. I had occasion to engage in a similar conflict in 1964, when the state Highway Commission (autonomous and arbitrary until it has become a statewide scandal) was determined to run a freeway through Golden Gate Park Panhandle—a beautiful approach to the city's largest and most famous park—a park that graced and gave play and open green area to a heavily populated section of the city. It was only a small group of citizens that initially opposed this move, and many who were aware of the strength and resources of the interests favoring the freeway plan thought they didn't have the hope of a chance. But more and more came in to help the little band of conservationists. The *Chronicle* backed the campaign, and called a big rally in the park's Polo Grounds. Even I was called on to come and sing my song "Little Boxes," which I was glad to do. But just before the rally I also wrote "The Cement Octopus," and I sang that, too. At this writing, after several years, the Panhandle still remains a park. A year after the first victory, a little group of us met in the Panhandle in the rain to celebrate and plant a small new tree there among the giant conifers and eucalyptus that had been planted by John MacLaren, who created the park in the preceding century.

And talents and dedication of all kinds rally in defense of the bay. The Bay Area Photographers do a whole season's show on the bay, and contribute prints to the Save the Bay Committee. Junk sculptors construct art works on the east bayshore flats, littered with flotsam, tires, cans, and other trash, as a sort of mute plea for something creative, some joy for the eye, to come out of this senseless ravaging of the bay. The songwriters sing of San Francisco, and the patriarch of folksingers, Pete Seeger, in his Columbia album devoted to conservation (the great conservationist Justice Douglas does the album notes for "God Bless the Grass") sings our song, "Seventy Miles."

What's that stinky creek out there,
Down behind the slum's back stair,
Sludgy puddle, sad and gray?
Why, man, that's San Francisco Bay!

Chorus:
Seventy miles of wind and spray,
Seventy miles of water.
Seventy miles of open bay—
It's a garbage dump.

Big Solano and the Montecell',
Ferry boats, I knew them well,
Creak and groan in their muddy graves,
Remembering San Francisco Bay.

(Chorus)

Joe Ortega and the Spanish crew,
Sailed across the ocean blue,
Came into this mighty Bay.
Stood on the decks and cried, "Ole!"

(Chorus)

Fill it there, fill it here,
Docks and tidelands disappear,
Shaky houses on the quakey ground,
The builder, he's Las Vegas bound.

(Chorus)

"Dump the garbage in the Bay?"
City fathers say, "Okay.
When cries of anguish fill the air,
We'll be off on the Riviere."

Chorus:
Seventy miles of wind and spray,
Seventy miles of water.
Seventy miles of open bay—
It's a garbage dump.

We are working out here to guarantee that this song becomes obsolete—a curio.

The Prudhoe Bay oil strike of 1968 represents a tremendous threat to the Alaskan environment according to Barry Weisberg (1944-), a free lance writer. It represents a threat, first, because no industry devastates its surroundings so completely as does oil; second, because the ecosystems of the Arctic are fragile and quickly disrupted by industrial intrusion. Weisberg sees the headlong rush for oil development in Alaska as an example of what Kenneth Boulding (page 307) calls "the cowboy economy," in which an apparently limitless earth is recklessly exploited without regard for the future, the public good, or the ecology of the region and its inhabitants. Weisberg then expands his criticism of the Alaskan oil boom into an indictment of the oil industry, its privileged and protected economic status, and its nationwide and worldwide, direct and indirect, contributions to environmental deterioration: air pollution from the internal combustion engine, oil spills on the earth's waters and shores, a car-oriented society with resultant urban sprawl, destruction of the landscape for freeways, and the like. In short, ". . . the oil industry, virtually a world government, presides over an economy organized toward the destruction of life."

ALASKA—THE ECOLOGY OF OIL

by Barry Weisberg

Americans would like to believe that the sins of Manifest Destiny are buried in the past, that the slaughter of the Indians and the extinction of the buffalo are but regretful memories, the stuff of history. But today in Alaska, this history is alive. There, drawn by the vast reservoir of oil discovered on the North Slope of the arctic coast, the awesome forces of American industry have assembled to re-enact the ruinous plunder of the great frontier.

To look at Alaska today is to return to a time when our waters ran pure, our landscape was unmarred by the oil derrick and the corner gas station, and the buffalo still roamed the open plains—Texas of 50 years ago or California before the turn of the century. What we are seeing in Alaska is a vivid compression of the past beauty and present devastation of the entire American environment.

To the popular mind Alaska—with 586,400 square miles, an area as large as California, Texas and Montana together—seems a vast, forbidding wasteland. In fact it is a land of incomparable beauty and resource, boasting endless cascades of timber, immortal rivers, mammoth glaciers; unbounded plains of caribou, grizzly bear, polar bear, and wolf; animals and plants unknown to most men. The amenities of clean air, water, and pristine habitation are unrivaled anywhere else in the world.

Alaska's antiquity can be discovered in the immense solitude of her mountains. Three great ranges transverse this land, many glaciated and silent beneath the ages-old mantle of snow and ice. Contrary to the usual image, during several months of the year Alaska is laden with brilliant poppies, roses, wild flowers, and vast continuous multi-colored fields. Hardly a bleak and uninviting world!

Alaska has more coastline (34,000 miles) than all the other coastal states combined. There is potential here for an estuary agriculture that could feed millions. There is more timber, water, and copper in Alaska than in all the rest of the United States combined. And, it appears, more oil.

On September 10, 1969, the corporate oil hustlers of the world descended on Anchorage for an unprecedented geological lottery in which they shelled out nearly a billion dollars—as much as $28,000 an acre—for a chance to exploit an oil field that may well turn out to be comparable to the massive field in the Middle East.

This peak price of $28,000 an acre was a respectable increase over the two cents an acre for which the U.S. purchased Alaska from Russia in 1867. As every school child knows, the area was originally meant to be kept as an icebox or a folly. But the discovery of gold at the turn of the century gave Alaska territorial status; in 1958, after many years of struggle by its white citizens, Alaska was granted statehood. Gold, fisheries and Government have brought with them the unwanted burdens of absentee landlordism. But it was not until early 1968, when oil was discovered around Prudhoe Bay on the northern arctic coast, that Alaska learned what real outside intervention was all about.

In 1965, '66, and '67, four major companies—Atlantic Richfield Company (Arco), British Petroleum (BP), Humble, a Jersey Standard subsidiary, and Sinclair—leased acreage for oil exploration on the North Slope, paying a total of 12 million dollars for leases now worth upwards of two *billion* dollars—or more than 150 times as much.

On February 16, 1968, Atlantic Richfield announced that its Prudhoe Bay No. 1 drilling rig, located two miles from the shores of the Arctic Ocean, had struck both oil and gas. Four months later, Arco's Sag River No. 1 rig, several miles to the southeast, struck oil.

In less than six months, a wilderness area the size of Massachusetts had been opened up to rapid development. In that time millions of pounds of equipment, fuel oil, pre-fabricated buildings, dynamite, people and food were flown in. Hundreds of miles of seismic lines had been run across the tundra, leaving permanent scars. And in a dramatic preview of the ecological disasters to come, a winter road was cut across the Alaskan wilderness to link Fairbanks with the Slope. The road, which was open for one month before it turned into the longest manmade swamp in the world, was officially named the Walter J. Hickel Highway.

Any objections to this "boom" raised by conservationists and others whom oil men find unaccountably superstitious about Industrial Progress, received a ready answer:

> *"There is oil out there.*
> *Somebody has got to get it out.*
> *You may not believe this,*
> *but it will be good for your town,*
> *good for the people."*

These particular lines were spoken by James Stewart in a movie (made long before the Santa Barbara oil catastrophe) about the world's first offshore drilling rig. But the same story has been given throughout history to every town and nation into whose land the oil industry has dug its iron claws: black gold will bring progress and prosperity.

Yet the Alaskan experience raises fundamental questions about this whole "development" process and the profit-oriented exploitation of resources—questions about the proper rate, purposes and forms of development, about who controls and benefits from it and by what right, and who really pays the price—questions about the heavy costs to life that do not show up on oil company balance sheets. These are the reservations that are obscured by the clichés of Progress, and overwhelmed by the euphoria of an economic "boom."

The $900 million paid to the State of Alsaka at the September 10th oil lease auction was touted as a munificent offering on the part of the oil companies. In fact it was only a fraction of the land's actual value. If prior lease sales are any indication, present value of this acreage is closer to $5 billion. Long-range value may soar as high as $50 billion within a decade. And while the state is now slated to receive a small share of ongoing revenues—a 12.5 per cent royalty and a 4 per cent severance tax—calculations by Gregg Erickson, a University of Alaska resource economist, indicate that the "State's severance tax/royalty can be raised to the vicinity of 85 to 90 per cent and still leave the oil companies a better than 10 per cent rate of return." The oil industries justify their profit by referring to the great risks they are taking in Alaska. But their claims are not terribly convincing. The journal *Oil Week*, in a much quoted statement, estimated that only five to ten billion barrels of oil were located in the Prudhoe Bay area. Yet over 50 per cent of the Alaskan geology is acknowledged to lend itself to anticlines, or oil-bearing structures. Interior Secretary Hickel, a well known partisan of oil, himself put the Alaskan reserves at about 100 billion barrels. Without question the Alaskan find will compare to, and likely dwarf, our primary domestic source today, the 30 billion-barrel East Texas find of the 1930's. Until Alaska, only 118 billion barrels of oil had been found in all of North America in the last 110 years.

This tremendous oil strike opens up the most perilous prospects for the Alaskan eco-system. No other industry could pose such a comprehensive threat to the wilderness environment of Alaska. No other industry can amass such large amounts of capital, or is so highly favored by tax laws. No other industry affects its environs as completely as does oil—its exploration, extraction, and transportation.

The significance of Alaskan oil development extends far beyond Alaska itself, carrying grave implications for our ecological well-being in its broadest sense, from the ongoing eco-catastrophe of our decaying, choked, polluted cities to the severe distortion in the allocation of basic global resources that American power imposes on the world. But to fully grasp these wider implications, one must first consider what the oil development means for ecology in the narrower sense, in the wilderness environment where the oil was found.

If oil is a uniquely devastating ecological enemy, Alaska is also a uniquely vulnerable victim. Until a very few years ago, Alaska remained essentially untouched by technological civilization. All of the organisms within its vast eco-system worked in complex and delicate symbiotic relations, species having survived and adjusted in accord with their ability to achieve symbiosis. Modern industrial society, on the other hand, works toward individuation and competition, toward conflict and instability. In the extreme but relatively stable and regular conditions of the arctic, the web of life-supporting relationships depends on the slimmest margins of sustenance. The slender food chains and parsimonious life-cycles afford little tolerance for disruptions in the pattern of balance. The slightest manipulation of the life support system, the alteration of a bird migration, the pollution of a river, the noise of an airplane, all have incalculable unanticipated consequences. That is what makes this unique and irreplaceable eco-system so utterly fragile and so vulnerable to the careless intrusions of industrial man.

In natural systems, the discarded and unused substance of one organism becomes the energy of another. With our consumption cult and profit-oriented technology, we seek to abrogate that rule, manufacturing and depositing waste with abandon. Industrial man's mania for waste is particularly disastrous on the Alaskan tundra, where debris survives intact longer than it does any other place in the world. Orange peels last for months, paper for years, wood scraps for decades; metal or plastic is practically immortal. The reason for this longevity is that arctic eco-systems are not prone to "bio-grade," i.e., to decompose matter. Because of the extremely slow decomposition rate, and the slow healing capacity of the mat of vegetative cover called tundra, the littering and desecration that normally take years in other parts of the world can happen almost overnight in the arctic.

Damage to the tundra is irreversible. This blanket of surface vegetation is a protective covering that insulates the deep layer of permafrost below. The permafrost, a mass of gravel, ice and mud that begins about a foot beneath the surface and extends downward a thousand feet or more, remains frozen throughout the year, providing a solid ground beneath the tundra. But when the cover is stripped away, the permafrost melts, leaving an open, unhealing wound of mud, slush and water that tends to drain away, undermining the stability of large areas of the surrounding earth.

The record of Alaskan "development" is written clearly in the tundra. Scars gouged by bulldozers 15 years ago remain distinct today. At Point Barrow and Amchitka and at the abandoned Naval Petroleum Reserve number four, one can see miles upon miles of oil barrels, wrecked airplanes and autos, Quonset huts and undistinguished junk—most if not all government donated. This is not merely an aesthetic problem. "Even now, 25 years later," says one observer, "many men who long ago left the Arctic still kill wildlife by the partially empty fuel drums they left behind. If conditions are right, they may wipe out an acre or two, or with luck, a whole lake."

But if the oil barrel in Alaska seems to some an ominous talisman foretelling environmental disaster, it is welcomed by others almost as an adornment. Colonel E. L. Hardin, chief of the Army corps of engineers in Alaska, says cheerfully, "The fifty-five gallon oil drum is the new state flower of Alaska." And the depth of the oil industry's concern for the environment, as well as the extent of its designs on it, is captured nicely in the response of one executive upset by the damage done by company equipment: "If we go on like that we won't have the remotest chance of getting into the wildlife range." However, rumors of clandestine explorations in the arctic national wildlife refuge abound.

There is little in the record of the oil development so far to inspire confidence in the future. The Hickel Highway fiasco—Hickel's last official act as governor—was only a hint of disasters to come. The road was to provide access to the Slope in winter when shipping is blocked and supplies must be flown in. It was a risky project at best, but the route was laid in what were obviously the worst areas, those with soils having the highest ice concentrations. When the ice broke up as summer approached, the permafrost melted and water from the adjacent land poured onto the roadbed, where it remains today. Hickel's response to critics was: "So they've scarred the tundra. That's one road, 12 feet wide, in an area as big as the state of California."

Years ago the extraction of any resource meant the establishment of a technological enclave, isolated for the most part from its surroundings. That at least was the model. Today oil extraction brings with it a whole supporting complex of advanced technology—"advanced" indicating that it is less restricted by or adapted to the natural environment and more able to impose its imperatives on the landscape, leaving it to nature to attempt to restore the ecological balance.

The primary challenge to the industry in this respect is transporting the oil to market once it is brought up from the ground. Roughly half of what the industry will spend exploiting Alaskan oil will go to providing transportation for it.

Plans are well underway for the construction of the Trans Alaska Pipeline Systems (TAPS). This $900 million pipeline will run from the North Slope some 800 miles across the Alaskan interior down to Valdez, an ice-free port on the Pacific southern coast. Described by the contractors as the "largest single construction feat in the free world," the pipeline will eventually transport some two million barrels of oil per day, at a temperature of 150°-170°. If placed underground, the builders admit that the line would melt all permafrost within a 25-foot radius. The actual pipeline trunk involves some 20,000 acres of land, but the roots necessary to support the venture will require another 7 to 9 million acres, accommodating 5 to 12 pumping stations, several landing fields, camp and administrative sites, microwave stations and access roads to the pipeline.

As noted in testimony given before the Department of Interior hearing in Fairbanks last August, "The construction and operation of a large, buried, hot [the oil must be heated to flow freely] pipeline in permafrost regions has never been done anywhere in the world." Of the many elements of arctic development, none has as great a potential for gross disturbance of the entire eco-system as does this pipeline. Laid upon a ten-foot bed of gravel (gravel taken from river beds, thus upsetting spawning and other cycles), it will almost certainly generate vast problems for soil stability in the permafrost—both because any intervention is hazardous to the delicate balance and because the varying ice contents of the soil require differing specifications for construction. The dangers of erosion, subsidence and stress to the surrounding environs are critical. Animals rely upon the vegetative cover for food and oxygen. To upset that balance is to intervene in the life-supporting processes of the entire biological chain of the arctic.

Wildlife patterns can be disturbed in an infinite number of ways. The mere physical obstruction of the pipeline itself would constitute a perilous barrier to the region's 400,000 caribou, blocking the migrations that are an integral and inescapable part of their life. To interfere with this ageless process invites a repetition of the fate of the buffalo.

This is only the beginning. Alongside the pipeline, the so-called "corridor concept" of development which TAPS is encouraging will string roads, railways, material storage centers and small settlements—and thus the inevitable forms of sprawl which follow such corridors. To talk about the pipeline, then, is to talk about an 800-mile strip of development, gross disturbance of eco-systems, and the basic interference with many life-giving cycles of the arctic.

On top of TAPS, the eagerness to get the crude petroleum out of Alaska to market has spawned the legendary voyage of the *Manhattan*, the 115,000-deadweight-ton super tanker that successfully passed through the arctic ice pack from the east coast of the United States to Point Barrow, Alaska, just west of the Prudhoe Bay oil area. Although the oil companies cited it as one of the "risks" of Alaskan development, they had already ordered eight more gigantic tankers before the first journey was completed. More than 1000 feet long, the *Manhattan* is able to crush its way through 40-foot-thick arctic ice. And orders are in for oil tankers three times as large. These technological behemoths—which in the course of their normal operations spew oil slick bilge and exhaust wastes in their wake—will cut a path of major disruption through more than a thousand miles of the arctic. As David Hickok, associate director of the Federal Field Commission for Development Planning in Alaska, notes, with such massive ocean-going ventures already in the works, there still exists an almost complete "lack of research and investigation in arctic waters on oil pollution, coastal processes, phytoplankton, marine fisheries and mammal populations, and on programs for the development of new technologies for port facilities in the arctic. All of these are prerequisite matters for governmental attention brought in focus by the voyage of the *Manhattan* and the granting of off-shore exploratory drilling permits on the continental shelf. . . ."

Over and over the oil industry ends up repeating, "No one could reasonably have expected": the million ton spill on the Delaware beaches; the splitting open of the Ocean Eagle in the San Juan Harbor; the collision which poured 30,000 gallons into the waters off the Cape of Good Hope; or the spill from the Torrey Canyon "whose captain ran her onto a well-marked granite reef off England in broad daylight, causing the biggest shipwreck and oil pollution ever." Or Santa Barbara! Such "unanticipated hazards" mark the operations of the petroleum industry daily. And to add to the peril, the petro-chemical industry is considering transporting pesticides (a by-product of crude petroleum) in similar large tankers. We are told that if the Torrey Canyon had been carrying pesticides rather than oil, the effect of such a shock could have abruptly terminated the production of oxygen by photosynthesis in the entire North Sea.

The gap between our ability to devastate and our ability to heal is enormous. In Santa Barbara, the highly complex equipment dedicated to pumping oil out of the ground contrasts sharply with the technology used to clean it up (i.e., hay spread across the sands), and in cases like the Santa Barbara Channel or the Hickel Highway, the damage is permanent, beyond repair in the time of man. To guard against such disasters would require time and the development of new technology—costly tasks that bring no profit to the industries involved.

In Alaska (where costs are high even when corners are cut), ecological precautions are certainly not allowed to interfere with profits. This assessment emerged from Senate hearings on Alaska: "Very frankly, in recent weeks, the committee [Senate Committee on Interior and Insular Affairs] has received a number of disturbing reports that present limitations on personnel and funding make it highly unlikely that proper environmental, conservation and safety control in connection with activities now underway or proposed will be fulfilled. I am hopeful that these reports are not true." It is a fleeting hope; anyone at all familiar with Alaska knows it is a futile one.

In Alaska today we are playing recklessly with forces which affect the entire planet. The arctic ice pack, for example, is perhaps the single most important land mass in determining global weather. It is possible that our interference with arctic heat patterns in the ice pack and the ocean (through oil explorations and transport) could upset basic weather balances affecting the height of the world's oceans, the amounts of rainfall, and other interdependent climatic functions.

There are all the symptoms of fatal pride in our tampering with these great harmonies. Thomas Kelly, the state official who presided over the big Alaska lease sale, was surely moved by hubris when he proclaimed, "To say that it is tundra today and should be tundra forever when tundra has no economic value doesn't make sense." Ted Stevens, appointed U.S. Senator in 1968 by then Governor Hickel, outdid Kelly in a speech before a meeting of the American Association for the Advancement of Science in Fairbanks last August. Stevens delivered a searing attack upon the ecologists who had come to discuss the oil development. After deriding out-of-state visitors as carpet-bag conservationists, he pulled out a dictionary and referred to the definition of ecology: "Ecology deals with the relationships between living organisms." "But," exclaimed the senator, "there are no living organisms on the North Slope."

Among the living organisms in Alaska which state officials would rather not think about are the native Eskimos, Aleuts and Indians, whose land the U.S. "bought" from Russia a century ago, and who still make up a sixth of Alaska's 272,000 population. According to the Statehood Act of 1958, 140 million acres of land were to be returned to the natives over a 25-year period. Years passed and the Alaskan native came to see clearly that the only way the white man could be made to live up to his 1958 "bargain" was through pressure. In 1966 the movement for native power coalesced into the Alaskan Federation of Natives, and their demands were formulated in the Alaska Native Land Claim bill.

In important respects, it was already too late. In 1964 the state, realizing that the North Slope was a potentially rich oil reserve, and that native pressure was mounting, applied to the Federal Bureau of Land Management (BLM) for the two million acres lying along the arctic coast in the Prudhoe Bay vicinity.

Although the land was a traditional hunting and fishing ground for the Eskimos, the state application claimed that it was free of aboriginal use and occupancy. The BLM then proceeded to publish notice of the state's intent in *Jessens Weekly*, a small mimeographed newspaper with irregular circulation. Thus, as Alaskan journalist Jane Bender comments, "The burden of proof was placed upon people who could not be expected to untangle the legal phraseology, who might not even have seen the notice in the first place, and whose knowledge of the far reaching consequences of that simple small print notice might be said to be minute."

The North Slope case was typical; the attitude of most white Alaskans is little better than colonial. So it is not surprising that the native claims have suffered continual erosion in the hands of all levels of government. In the first compromise the natives settled for 80 million acres; they were then forced down to 40 million acres. Walter Hickel now suggests 27 million and the governor, Keith Miller, suggests 13 million acres—out of the total of 365 million in the state—3.6 per cent of the land for a sixth of the population, when rightfully they own it all.

It is safe to assume that the empires of oil will wield their vast power to delay a native settlement. For if significant portions of the state were in the hands of the natives, the oil combine would have to deal with them rather than the state, and they are potentially much less willing accomplices in the rape of the land—as evidenced by their picketing at the lease sale.

Meanwhile there is increasing pressure to lift the freeze on land giveaways to state and private interests that was imposed by former Interior Secretary Udall pending a settlement of the native claims. For example, the TAPS pipeline by law must secure a lifting of the freeze in order to proceed. However, as a *New York Times* editorial reports, "There is good reason to believe that preliminary work on the right of way has been started, without benefit of permit or of law."

Of course the oil industry has every reason to be confident that the government will smooth the way for it in Alaska. The $20 billion a year industry is famous for its unsurpassed political and economic power in America. Its lobbying muscle in Congress is legendary. It enjoys the lowest effective tax rate of any U.S. industry (seven per cent for the 23 largest companies). The oil depletion allowance is a prime symbol of corporate privilege, yet tax reform on it has been held to a meaningless reduction of a few percentage points. (Because of existing restrictions most large companies use only 24 per cent out of the current 27 1/2 per cent allowance anyway.

An industry that has been able (notably through the Rockefeller—Standard Oil complex) to treat the U.S. State Department as a subsidiary headquarters, and at whose bidding America brings down sovereign governments (as in Iran), should not expect to have much trouble making its way in Alaska. Still, as extra insurance against public clamor, Atlantic-Richfield (which made the first North Slope strike) and other companies are investing large sums in advertising, and even in conservation groups, in an effort to control public awareness of key development issues.

The industry had little difficulty getting someone sympathetic to and familiar with their Alaska problems into the key Interior Department post. It is generally accepted in Washington that Atlantic Richfield's chairman, Robert Anderson (who as Secretary of the Navy encouraged the opening of Alaskan lands to private development), was most responsible for President Nixon's appointment of Hickel to Interior. Certainly Hickel, with his celebrated oil connections, and his financial interests in the copper of the Brooks Mountain Range and the Yukon River Delta—another potential oil reserve—was not appointed for objectivity, nor for public relations finesse on conservation, given his plans to "build a Fifth Avenue on the tundra."

But the critical importance of the industry's political power in Alaskan development is most clearly revealed in an enormous irony: it is only the industry's ability to use government regulation to rig the American oil market that assures the profitability of exploiting Alaskan oil in the first place. It is, in other words, the artificial overcharges imposed on consumers, rather than its intrinsic economic profitability, that is underwriting the current rapid development of Alaskan oil and the environmental disruption that goes with it.

The rigging mechanism involved is the Oil Import Quota Program, which is conservatively estimated to cost the American consumer $4 billion a year. Production costs for oil from Texas and other domestic U.S. sources are far higher than those from rich, easily worked foreign reserves such as Venezuela or the Middle East. Even when the cost of transporting foreign oil to the U.S. is added on, its price at the point of delivery here is little more than 60 per cent that of domestically produced oil. The price of a barrel of comparable crude oil delivered to Philadelphia is about $2.25 for Mid-East compared to $3.75 for domestic. Allowing foreign oil to compete freely in the U.S. would drive all but the cheapest and most efficiently produced domestic oil out of the market. Instead, the government obligingly sets severe limits (presently 21 per cent of domestic consumption) on the amount of oil that can be imported. All oil in the U.S. is then sold to the consumer at the high prices of domestic oil.

Alaskan oil, according to current cost projections, is expected to be too expensive in production and transportation to compete on the world market with Middle Eastern and other oil. However Alaska is a state, and Alaskan oil is therefore domestic American oil. Thus, like the privileged petroleum of Texas, it can be sold in unlimited quantities at the artificially inflated prices of the U.S. market.

In a sense the current Alaskan development can be considered an elaborate economic charade, in that profits come from the industry's power to levy exorbitant prices against the consumer, rather than from the normal proceedings of business. We would all be better off, in fact, just to pay the extra money directly into the corporate treasuries, plus a little bit to the State of Alaska in lieu of royalties, and have the companies leave the Alaskan environment alone.

It is quite possible, of course, that as time goes by and the impact of the magnitude of Alaska's reserves is felt, the cost of Alaskan oil will come down to an economically competitive level. The fact remains, however, that the financially riskless, headlong development we are seeing depends on the industry's ability to supersede the laws of the marketplace and to have its way in the affairs of men.

While there are stirrings of opposition to what is happening in Alaska, they are largely isolated and, consequently, impotent. Many legislators told me privately that they thought the oil lease sale should have been postponed but were afraid to say so publicly for fear of losing re-election. The Federal Field Commission for Development Planning in Alaska has repeatedly advocated more planning, more care, and a slower pace—to no avail. One legislator said, "The trouble is that the state government has said so little about the problems of the North Slope that I can't tell if they are ignorant, unconcerned or are withholding information for other reasons."

In Alaska, as elsewhere, the tremendous power of the oil industry over social development grows not only from its impressive ability to dictate government policy, but also from the extent to which patterns of development are set autonomously by the "private" operations of industry. It is the general void of public policy that gives industry a free hand to shape the future in terms of its private priorities, unchecked by public interest or authority. When government does intervene (usually under the influence of industry anyway), it is merely responding to the reality that industry has created.

There is virtually no public policy governing the pattern of social growth on the Alaskan frontier; there is no involvement by the people of Alaska or of the rest of the country, no informal advocacy procedure by which to evaluate what the oil companies are doing. As a result, oil exploration and production proceed in Alaska without any projected land use plan, without any legislative priorities for growth. There exist no uniform codes for oil and mineral exploration, no systematic efforts toward the preservation of wildlife populations, nor any air or water quality standards to speak of.

This abdication of public discretion is not an accident. David Hickok complains, "Both industry and government are deliberately preventing the operation of a public forum until after the important decisions are made." The problem, then, is not that our current situation results from no planning, for clearly the oil companies have a very keen sense of plan and purpose. It is rather that the plans which do exist are created and executed without public scrutiny or control.

In just "going about its business," the oil industry will change the face of Alaska more thoroughly than all the volumes of hotly debated social legislation that give people the illusion of controlling their own destiny. Meetings have been summoned throughout Alaska by the Brookings Institution and the Stanford Research Institute to discuss the state's development plan. By invitation, participants will be discussing the future while oil is already determining it.

The coming of the oil empire to Alaska brings with it the vast support operations of railroads, airlines, communications networks, new towns, urban growth and the like. Requiring highly skilled labor, these operations will not draw primarily on either natives or local whites. Consequently, entire communities of technically skilled men will be brought into Alaska: 5000 on the Slope, 3000 for the pipeline, 1500 at Valdez, hundreds more administrative persons in Anchorage and Fairbanks—all of them requiring housing, food, and related services. What will happen when initial construction of the new industry is completed and they leave? What will be the effect on the economy when some 10,000 people who have been disbursing enormous amounts of capital pull out in one year? The initial boomtown profits to local businessmen and landholders will give way to the ghost towns that followed the gold rush. The oil rush economy does not build for posterity.

Already the landscape of Alaska is dominated by a crude mix of the worst Texas gulf coast and Southern California plasticity. Housing is composed almost entirely of imported pre-fabricated units or trailers. A ticky-tacky frontier bar atmosphere permeates every Alaskan town. Within a very short time oil has penetrated all aspects of the Alaskan economy. In terms of outright ownership, the industry is gobbling up local business interests at a rapid rate. The income of hotels, restaurants and airlines depends upon the oil companies. The universities are in their employment.

The economic reality of oil development in Alaska is that control of the revenue (perhaps $100 billion in the next decade) implies power over social development. The unchecked nature of this power must force us to reconsider our whole system of ownership and extraction of resources. For instance—to consider the minimal alternative—if this oil were to be developed by a quasi-public corporation (looking to the TVA as one example), the revenue involved could be employed to initiate environmental quality regulatory programs or the reconstruction of urban cores. The present arrangement abdicates what may turn out to be the only source of revenue the United States could ever have to begin to cope with urban pathology. We presently act as if the oil were ordained by geology just to serve the industry, as if the oil companies somehow owned the oil which they sell to the people of the world at such fabulous profits.

The oil finds in the arctic are tremendously important in themselves, but it is the dynamic they lock us into that sweeps out to affect the lives of every single person in the United States, and throughout most of the world. The results are simply taken for granted, without regard for the decisions that initiated them. The fact that oil development fixes the American landscape into transport corridors geared to accommodate cars, and only cars, is no small matter. For to insure the presence of the auto is to insure the persistence of current forms of urban sprawl. To pressure against cooperative modes of transit is to fix the shape of the city for tomorrow.

The technology of the automobile itself is obviously tightly meshed with that of oil. Cars presently consume upwards of two-fifths of all the crude oil produced. And it is clear that the environmental pollution that goes with the internal combustion engine and other petroleum-funded technology is not a mere matter of irritation or inconvenience. The head of the American Petroleum Institute's 300-member Committee on Air and Water Conservation continues to decry the "passion" and "emotionalism" which mark opposition to air pollution. He has said that "we can go along as we are now for another 10 to 15 years." But University of California Zoology Professor Kenneth E. F. Watt predicts, "It is now clear that air pollution concentrations are rising in California at a rate such that mass mortality incidents can be expected in specific areas, such as Long Beach, by the 1975-76 winter. The proportion of the population which will die in these incidents will at first equal, then exceed, that for the 1952 smog disaster." (Nearly 2500 Londoners died from the effects of smog during the Christmas season of that year.)

Already the children of Los Angeles are not allowed to "run, skip or jump" inside or outside on smog alert days, by order of the L.A. Board of Education and County Medical Association. If the oil and auto industries had spent a fraction of their advertising budgets on research for a smog-free engine, our air today might be safe for future generations.

Oil is at the core of the whole of American industry. Crude petroleum is the basis for the production of hydro-carbon feedstock and other basic petrochemical industries. Petrol is the stuff from which roads, paints, detergents, synthetic rubber, cosmetics, nylon and pesticides are made. From its powerful position at the center, the oil industry fuels, sustains and protects the economy of waste on which its profits are based. It defends and expands that economy's myriad patterns of devastation: the private auto, in use about one per cent of the time and junked at a rate of 12 million a year, usurps 50 per cent of the space in our overcrowded cities with its highways and parking requirements; the use of DDT and other less celebrated pesticides generates crop surpluses which are then withheld from the hungry at home and abroad; the ubiquitous plastic packaging is neither reusable nor decomposable, and it pollutes the air when burned. This is the technology with which we are "developing" Alaska and civilizing the world. This technology costs $11 billion annually in damage to private property from air pollution alone.

The headlong rush of Alaskan development is part of a momentum that completely contradicts our knowledge about the capacity of the earth to support us—namely, that the resources of the earth are fixed; that, rather than continuous growth merely to accommodate the increasingly false consumptive needs of an increasing number of people, growth must be directed to achieve very specific public priorities—priorities which are determined by the kind of life-styles which neighborhoods and regions determine are best for them. Limits must be set. Development as it now proceeds minimizes the alternatives open to people, increasing the uniformity and standardization of life. It locks us into patterns over which we have little knowledge or control.

It is not enough merely to slow down in Alaska, as a *New York Times* editorial of November 10 argued. Development as it proceeds on the North Slope, and on countless other frontiers of American industry, must be curtailed. Until such time as the American public has adequate time and information to evaluate and assess the total costs of industrial development to all the people affected by it, development and the myth of growth must be curtailed. Rational resource consumption and re-cycling alone would eliminate the need for any further oil extraction on the face of the earth.

While the population of the world is expected to double in 35 years, it will consume resources at not twice, but five times the present rate, producing a scarcity in food and fossil fuels that will be the major source of friction in the coming decade. This results directly from consumptive patterns generated by the United States. It is the disequilibrium between man and nature, not the biologic process of procreation, which is at the root of the population issue. To cope with population is first and foremost an issue of coping with the current American imperialist consumption of 70 per cent of the world's resources by less than 7 per cent of the world's population.

The largest single consumer of crude petroleum is the American military—those who are charged with defending this squandering of other people's resources. Alaska is key to their continued world supremacy. As America shifts in Southeast Asia and throughout the world to air power rather than ground forces, the military appetite for oil will grow and will seek stable sources. Walter Levy, known as the dean of U.S. oil experts, points out, "A world power which depends on potentially reluctant or hostile countries for food and fuel that must travel over highly vulnerable sea routes is by definition no world power." While we "own" major portions of Mid-East reserves already, the transport of this oil is in constant jeopardy, as the closing of the Suez in 1967 showed. And domestic production, aside from Alaska, is projected to fall behind consumption at an increasing rate in the next few years. Alaskan reserves will stabilize the strategic military supply of crude oil. Moreover, as America prepares for the rearmament of Japan to help police Asia, treaty negotiations are already being pursued in Washington to provide Japan with a stable oil flow from Alaska in exchange for military and trade arrangements.

What Americans must realize is that the destruction of our life support systems will not be halted through our individual refusal to drive cars or use pesticides. As is evidenced by the Alaskan oil rush, development no longer proceeds along enclave lines, but is comprehensive in impact and scope, so that conservation efforts which act to preserve wilderness enclaves as parks or wildlife refuges will in the end lose those areas to the all-inclusive effects of air pollution, noise, and pesticides. The oil industry, virtually a world government, presides over an economy organized toward the destruction of life. Its power must be broken, not merely circumvented. The avenues of oil must be reached at their point of production, not merely in our own individual use.

If anything is to be learned from Alaska, it is that it is time to stop. Domination through growth has mesmerized the American mind for so long that the suggestion of curtailing growth is unthinkable. Greek rationalism, the Roman engineering mentality, the Biblical injunction to conquer and subdue nature, the post-Enlightenment mystique about technical progress—all espouse development.

Yet the old myth that continued growth increases our control over the environment is now simply false. We are losing control. We are destroying the air we breathe, the water we drink, and the land we walk upon. And this is not an accident. It is rooted in the fundamental attitudes and practices of advanced industrial society. It is in part the logic of capitalism, but it is more than that; it is the very relationship we assume toward the natural world.

The talk about shifting from an economy of affluence, obsolescence, redundancy and waste to an economy that recognizes scarcity must yield practical proposals for a new economics. And these proposals must include the mandatory re-cycling of all natural resources; the mandatory production of only re-cyclable containers; the rationing of all natural resources—rationing to provide for sane limits on the amount of consumption as well as to equalize mechanisms for distribution.

Industrial processes must be rationed as to the amount of oxygen, water or minerals they can consume in production. These are no small matters, but they are only the basic parameters for what would be the beginning of a truly democratic policy for our life support systems. The "economy of death" must be replaced by an economy of life.

Continued rapid development such as that in Alaska can only work for the forces of exploitation and greed. Time must be had to examine and consider every aspect of the development process, to create a comprehensive democratically determined land use policy, to devise environmental regulatory agencies with adequate means of enforcement, to develop new forms of revenue sharing and community control over economic growth, to re-learn our inclinations toward nature and our relationship to people unlike ourselves. While this must happen in Alaska, it must also happen on a national and global level. For clearly the powers that shape the fate of Alaska are rooted in places far distant from that beautiful land.

We must slow down. We must come to enjoy the world gently, remembering that this fragile earth is more to be admired than used, more to be cherished than exploited. Alaska teaches us that there are men for whom this is impossible. They must be stopped. Not for their sake, but for ours.

One target of Barry Weisberg's attack upon the oil industry—the automobile—is probed in greater detail by Frank Graham. Jr., who also wrote "Mississippi Fish Kill" (page 153). Virtually everyone is already aware that the automobile is a leading cause of air pollution, but what may not be known is the extent to which automobile manufacturers have ignored or evaded the problem except when threatened by federal intervention or a successful competitor. (Graham also mentions several examples of what has been cynically called "eco-pornography," that is, the sort of advertising campaign which assures the public that the pollutants of business X are really harmless, or that oil wells are good for wildlife, or that the auto industry is committed to solving the smog problem.) Graham further considers less notorious smog producers such as jet planes and diesel engines; he examines the alternatives to the internal combustion engine, such as the electric or steam car; and he describes the actions of state and local government and citizens' groups in combatting air pollution. Nevertheless the seriousness of the threat is still not fully acknowledged, and, as Paul Ehrlich warns in the opening essay, time is running out.

THE INFERNAL SMOG MACHINE

by Frank Graham, Jr.

Mobil Oil Corporation recently issued a booklet called *A Primer on Air Pollution*. It tells us that we have a problem, which comes as no surprise. With hearty candor, it admits the culpability of that compact little bundle of pollutants which elsewhere has been called the "infernal combustion engine." The primer mentions possible alternatives, including the electric automobile, and even wishes the alternatives well ("Competition is the name of the game"). But since electric autos do not use gasoline, Mobil quickly dismisses them.

"We feel that voluntary action is best," Mobil says. "We're doing everything we can to make sure something happens: a car and fuel combination that's better than 95 percent pollution-free."

Ever since industry came around to admitting that there was a pollution crisis (which was not very long ago) it has subjected the public to a stream of similarly simplistic primers. Most of them are attractive. Most of them tell a plausible story. And all of them assure the reader that industry, left to its own methods, is rapidly clearing up whatever little mess it created in the first place.

Unfortunately, these beguiling booklets contribute to the breakdown in communications that let the pollution problem get out of hand in the first place. Nor is the situation getting better. Despite our current efforts to improve them, things may very well get worse.

Because of its omnipresence, the automobile has replaced all of the country's stationary sources—its smoking chimneys—as our most publicized air pollution problem. There are about 90 million autos on American highways today. Warren M.Dorn, of the Los Angeles Air Pollution Control Board, has sketched this machine in all its awesome multiplicity.

"The automobile, which is responsible for the emission to the Los Angeles atmosphere of 80 percent of the smog-causing hydrocarbons and 50 percent of the smog-causing oxides of nitrogen, is uniquely a creature of interstate commerce," Dorn said. "Its raw materials, its subassemblies, its finished products move in interstate commerce. Upon the well-being of the industry that produces it rests a good portion of the well-being of our economy. Its use has remade the appearance and the social structure of the United States. The waste products arising from its use now threaten the health, welfare, and comfort of people in communities from coast to coast."

It was in Los Angeles in 1950 that the automobile's contribution to urban smog was unveiled. Hydrocarbons, reacting with oxides of nitrogen under the influence of sunlight, were found to produce what we now call photochemical smog.

Because of Los Angeles' special climatic conditions, pollution became readily apparent there before other cities considered it a menace. Yet Los Angeles' concentration of automobiles is lower than that of many other American cities, including New York, Chicago, Detroit, Washington, and Philadelphia.

Cars and trucks crowd into cities, creating dense aerial sewers wherever they come together in nerve-jangling tie-ups; it is said that a mighty truck in New York City today averages only six miles per hour, while a *horse-drawn* truck in 1910 averaged eleven miles per hour on the same streets. An auto, idling in traffic jams, produces at least one billion airborne particles each second.

What are the principal sources—and components—of automobile pollution?

If a car is not equipped with smog-control devices, one out of every ten gallons of gas poured into its tank is returned to the atmosphere. This gallon is not only wasted (a drain on our pocketbooks and on our natural resources), but it appears in the air as pollution.

Ten percent of this offending gallon escapes as evaporation from the carburetor and fuel tank.

Twenty-five percent leaks out of the cylinders (between the pistons and cylinder walls) into the crankcase. Engineers refer to this leakage as "blow-by." These vapors, too, finally reach the open air.

Sixty-five percent of that gallon is pumped into the environment through the tailpipe. More than 200 different chemicals have been identified in auto exhausts, many of them created by sunlight acting on simpler compounds. In these exhaust emissions scientists find hydrocarbons and oxides of nitrogen. Here, too, they find carbon monoxide (that painless antidote for the suicide's ills). And here they find lead—derived from anti-knock additives—which is now causing so much concern.

As stationary sources of pollution are brought under control (most dramatically in Los Angeles), the automobile contributes an even greater share of pollution. The Connecticut Agricultural Experiment Station at New Haven has demonstrated how auto exhaust gases blanket the countryside, damaging farm crops miles from the nearest highway.

Naturalists have speculated that lead from auto exhausts may contribute to the buildup of environmental contaminants, and so interfere with the successful reproduction of such birds as the bald eagle. Auto exhausts certainly add to the urban pollution, which is now widely conceded to damage human health.

"The more we look for signs of pollution damage," says a U.S. Public Health Service official, "the more widespread we find it to be—whether our criterion be vegetation damage, materials deterioration, visibility decrease, or health effects."

Yet even the attention suddenly focused on this problem has not brought us much closer to solving it. Another Public Health Service statement, from a technical report, explains why: "The expected increase in the number of motor vehicles will far outweigh the partial reductions which can be expected from presently available crankcase and exhaust-control systems."

Los Angeles is nearly choked by smog; its climatic conditions act to press pollutants back down upon the city, rather than allowing them to escape over the mountains to the east. Yet one natural blessing that has helped Angelenos to survive is the immaculate air which comes to them from the Pacific Ocean. There are, obviously, no serious polluters to the west.

But once upon a time, in a simpler world, the American businessman dreamed of a day when every Chinese family would own its own car. Imagine the rubber tires, jacks, and gallons of gasoline to be sold to the stirring Asiatic giant! That dream has reappeared as a nightmare for pollution control experts on the West Coast. The vapors drifting across the Pacific from 200 million Chinese automobiles might very well convert Los Angeles County into a literal Death Valley.

A small but still significant amount of pollution is pumped into the environment by jet planes and diesel trucks. On takeoff, for instance, a four-engine jetliner releases 88 pounds of pollutants. A question now being pondered seriously by scientists is: Do jet contrails restrict the amount of sunlight reaching the ground, and so reduce the Earth's energy income? No one seems to know.

"I have never seen so much smoke come out of airplanes," an executive of Britain's Rolls-Royce Limited said after a visit to the United States. "That shouldn't be permitted. Our engines don't do that."

All of us are familiar with the odious plumes of diesel exhaust—composed mostly of solid particles of carbon—that pour from trucks. Now we learn that this smoke is almost wholly unnecessary. Most of the blame is laid on truck operators, who attempt to increase the power of their engines by increasing the amount of fuel that is fed to them.

"There is almost unanimous agreement among the experts that a well-designed engine, in good repair, using the proper fuel, will not emit visible smoke."

That was the testimony of Dr. P. H. Schweitzer, an expert on diesel engines, before a Senate subcommittee not long ago. "It is similarly accepted that when overfueled (overloaded) any diesel engine will smoke," Dr. Schweitzer went on. "As our passenger automobiles are overpowered, our trucks are generally underpowered. They do not have big enough engines."

Industry likewise discounts the effects of diesel emissions. "First," says Dr. William G. Agnew of the General Motors Research Laboratory, "there are too few diesel vehicles and too little diesel fuel consumed for diesel exhaust to contribute importantly to overall urban air pollution. So diesel exhaust is a localized problem generally of concern only within 100 feet or so of the individual vehicles.

"Second, no health effects are involved with diesel exhaust in normal urban environments, so the problem is primarily a matter of nuisance." (Detroit once tried to camouflage the discomforting stench from its fuming buses by mixing cheap perfume into the fuel.)

Still, the Department of Health, Education, and Welfare recently issued standards designed to regulate the smoke from diesel engines. They will go into effect with 1970 model trucks and buses. Even then, adds Dr. Agnew, diesel users could offset the control efforts of engine builders. Cheaper fuel, reduced maintenance, and producing more power than intended mean short-term cost savings—and more smoke. "It appears," says the General Motors expert, "that only legal enforcement of universal smoke-emission standards can eliminate the competitive economic advantages and bring about control." In the meantime, researchers in the GM laboratory are using the human nose to determine the acceptable "odor thresholds" for smelly diesel fumes.

But it is the automobile that has been the most aggravating source of pollution for a good many years. And the industry did its best to keep this fact a secret, just as it did its best to forget the number of mechanical defects in its products until the advent of Ralph Nader. If the auto industry has had a similar gadfly in the pollution field, the accolade must go to Kenneth Hahn of the Los Angeles County Board of Supervisors.

"We feel that the automobile industry has failed its responsibility to the American public and to the health of the community," Hahn says.

He speaks from considerable experience. Hahn began his crusade so long ago that he numbered the Packard Motor Company among his targets. Fortunately, this experience is amply documented in a file containing Hahn's correspondence with leaders in the auto industry.

Los Angeles, of course, was aware of the source of its smog as early as 1950. The industry paid no attention to the city's complaints. In 1953, Hahn finally wrote to the presidents of the major automakers, expressing the concern of many local citizens. On March 3rd, he received a reply from a Ford public relations man.

"The Ford engineering staff, although mindful that automobile engines produce exhaust gases, feels these waste vapors are dissipated in the atmosphere quickly and do not present an air pollution problem," the Ford man wrote to Hahn. "Therefore our research department has not conducted any experimental work aimed at totally eliminating these gases.

"The fine automotive powerplants which modern-day engineers design do not 'smoke.' Only aging engines subjected to improper care and maintenance burn oil.

"To date, the need for a device which will more effectively reduce exhaust vapors has not been established."

Hahn persisted. At intervals he shot off letters to the industry's Big Three. Invariably the answers were unsatisfactory. A General Motors vice-president (at least Hahn's correspondents were climbing the corporate ladder) assured him that GM was working hard to reduce pollutants in auto emissions.

"I would be less than candid, however," the Detroit executive wrote, "if I failed to point out that we believe that these changes on automotive vehicles alone will have a very limited influence on the total smog problem in Los Angeles."

For the next several years the correspondence lagged. Ford Motor Company, for instance, seemed to lose interest in Hahn. Intent on developing a brand new car, its executives were wrapped up in a long correspondence with Marianne Moore, during which they hoped the poet would come up with a name for their product. (After Ford pulled the rug out from under Miss Moore by naming its new car the Edsel, the public pulled the rug out from under the car.)

In 1958 Hahn was still hammering away. "I have gained the impression," he told the Chrysler people, "that the automobile industry considers that air pollution is not as important to the industry as the new styling, grille design, more horsepower, or any other accessory."

By 1962 Hahn was receiving replies directly from John F. Gordon, then president of General Motors. Gordon complained that the California Motor Vehicle Pollution Control Board had not seen fit to approve any GM control devices; they had been turned down as ineffectual.

A year later Hahn wrote more forcefully to Gordon:

"Your company has done much to improve the performance and appearance of its products with one notable exception. Despite the great advancements in styling and horsepower, the smog-forming capability of the exhaust of your 1963 models is the same or more than the 1953 products."

Later that year a Ford official began his reply to Hahn in a manner that might have been considered a put-on by a man who had had less experience in dealing with the high and the mighty.

"Thank you for your letter of July 30, 1963," the Ford man wrote. "We are gratified that you continue to share in our concern with the Los Angeles smog problem."

The Ford man went on to blame the whole thing on worn-out cars. His letter seemed to suggest that the problem might be solved to everybody's satisfaction if every Angeleno rushed out and bought a spanking new Ford.

Unfortunately, Hahn and other California officials were responsible for most of the pressure on auto manufacturers during those years. No one else was especially interested, except perhaps then Secretary of Health, Education, and Welfare Abraham A. Ribicoff, who finally exacted a promise from the industry to equip 1963 and later models with devices to control blow-by emissions (hydrocarbons) from the crankcase.

California established its Motor Vehicle Pollution Control Board in 1960. A year later, the industry equipped all cars that were to be sold in California with devices to control blow-by emissions from crankcases. These devices were not available on cars sold in other states until 1963. By 1966 California was also able to require devices to control exhaust emissions on all new cars sold in the state.

Industry likes to pat itself on the back for its "voluntary" installation of control devices. Nevertheless, California officials claim that the manufacturers began to develop such devices only when it became clear that outside firms might market them first.

"Development of suitable means to control vehicle exhaust was a slow process," says John Maga, chief of the Bureau of Air Sanitation of California's Department of Public Health. "Initially there was widespread interest in device development. The lure of a ready-made market provided the incentive for several companies to enter the field. But when several catalytic and afterburner mufflers were certified, the automobile manufacturers introduced devices of their own. The new-car market was eliminated insofar as independently developed devices were concerned. The withdrawal of these companies from competition has removed much of the stimulus for further device development by automobile manufacturers."

There remains considerable discontent in California. The antipollution devices, in most cases, have been installed only on new cars. Eighty percent of the state's cars (about eight million of them) still have no such equipment. Moreover, the existing devices often fail to function properly. Industry wants to see stricter inspection procedures set up to detect the faulty devices.

"The auto industry taking this position is understandable," says Louis J. Bintz of the automotive engineering department of the Automobile Club of Southern California, "since it is only natural for them not to wish to assume the responsibility for the deterioration of these systems but rather to make it the responsibility of the motoring public . . . But maintenance has very little, if any, effect on air pollution emissions."

The auto industry, which first balked at California's stiff emissions standards, later balked at installing exhaust control devices on cars sold elsewhere in the nation. It was industry's position that California's problem was unique. To equip all cars with devices, one industry newspaper claimed, was a "billion-dollar hoax."

Yet the federal government has ordered emission control devices on all new cars. Its sensible position is that the new regulations do not, as industry contends, "discriminate against smog-free localities." Most cars are operated in polluted areas, and almost all cars, even those from the smallest towns, eventually enter cities. The quicker these regulations go into effect, the sooner all cars on the road will be equipped with such devices. As it is, the older cars that lack them will still be with us for many years.

But what about the industry's exhaust devices? They are designed to control only hydrocarbons and carbon monoxide. Critics claim engines so equipped may take in more air. The result, they say, is that atmospheric oxygen and nitrogen combine to form even greater amounts of oxides of nitrogen.

"We do have indications that bronchial damage occurs in humans exposed to a few parts per million of nitrogen dioxide," says Edward A. Schuck of the statewide air pollution research center of the University of California.

"And we do know for a fact that the nitrogen dioxide concentrations, which are now characteristic of the Los Angeles atmosphere, cause measurable decreases in the growth of vegetation. We also know that any increase in the nitrogen dioxide concentration will result in further visibility reduction because of the brownish-red color of the gas."

So auto pollutants still stream into the air, and Los Angeles is stuck with its smog. A waggish football coach recently explained to a reporter how his team trains for a game in Los Angeles.

"First we practice with one nostril taped," he said. "Then we have an assistant coach smoke a big cigar in the bus on the way to the stadium."

The Los Angeles smog, incredibly, is killing a wilderness forest 60 miles from the city. In the popular San Gorgonio Wilderness in the San Bernardino National Forest, 25,000 acres of ponderosa pine have been afflicted by blown-in smog disease.

Foresters say the ozone in the smog destroys chlorophyll in the needles and speeds aging of the trees. Death for the yellowing ponderosa is only a matter of time under continued exposure to the foul air.

Smog damage is even appearing in other national forests in California. Researchers hope to find individual trees which are smog-resistant, and determine how smog affects other western conifers. But, they emphasize, there will be no cure, only prevention—of air pollution.

St. Louis, meanwhile, has moved to combat auto pollution by ordering a special force of police officers to ticket drivers whose cars lay down a smoke screen. Since many of the most damaging pollutants are scarcely visible, especially in cold weather, this measure has obvious limitations. Conversely, New York City has considered banning automobiles from certain congested areas when carbon monoxide fumes approach the danger level.

In this consideration, never invoked, lies the germ of a still more revolutionary solution to the air pollution problem. More and more, one hears the ultimate reasoning: The internal-combustion engine will always throw off pollutants; and, with the constant increase in the number of cars on our highways, even a great reduction in each car's pollutants cannot prevent us from putting an insupportable burden on the environment. Thus, the automobile may become a luxury that cities cannot afford.

"The day may come," former Secretary of Health, Education, and Welfare John W. Gardner has said, "when we may have to trade convenience for survival."

What *is* a feasible alternative to the internal-combustion engine? The answer, from many quarters, is the electric car. At present, forklifts and other small vehicles operate on batteries in areas (such as warehouses) where fumes would be intolerable. These, however, are lead-acid batteries. Their power is not sufficient for high-speed highway travel.

An electric-powered all-purpose automobile is still many years away. Batteries run down quickly and, therefore, limit a car's range. (They also limit a car's speed, but this might not be catastrophic.) However, many air pollution experts now suggest that cities look into the possibility of applying for federal funds to develop electric taxis and buses. Such uses clearly lie within the capability of battery-powered engines.

Early autos, of course, *were* powered by batteries. The internal-combustion engine finally proved more efficient, and electric cars disappeared. Technologists lost interest in them, so that today, according to *Fortune* magazine, "The batteries required to drive a Volkswagen, with equal performance to that of its gasoline model, would fill a second Volkswagen hooked on behind as a battery carrier."

Now scientists are looking into a variety of batteries which eventually may rival the internal-combustion engine as a means of power. One hears of zinc-air batteries, sodium-sulfur batteries, and lithium-chlorine batteries. Perhaps the most feasible of all is the fuel cell, which can be refueled like a gasoline engine. The fuel cell, however, may be 25 years away.

The steam-propelled car—rarely seen for half a century—also has its advocates. In two days of recent Senate hearings, proponents claimed that steam cars would virtually eliminate automotive air pollution, that they would be superior in quality and performance to internal-combustion vehicles while producing 95 to 98 percent less harmful emissions.

Reporting on the hearings, *The National Observer* said "none of the witnesses was more enthusiastic about the possibilities of steam than consumer crusader Ralph Nader." Nader said he would do "everything I can" to persuade the new federal Department of Transportation to finance construction of a prototype demonstration steam car. "The government is spending hundreds of millions of dollars to subsidize the supersonic transport plane," Nader said, "so why shouldn't it be willing to spend $5 million on a project that could enhance the health of more than 200 million people."

Numbered among the scoffers at the Senate session were spokesmen for Ford and General Motors. But Dr. Robert U. Ayres, who researched the subject under a Ford Foundation grant, said that within three years, production lines could be turning out cheaper-to-run steam roadsters at lower costs than conventional autos. And New York Congressman Richard L. Ottinger promised to introduce a bill to require the federal government to buy steam or electric cars "wherever feasible" to create a market.

Whatever becomes of steam and electric cars, they may have already performed their greatest service in looming as a bugaboo before the automobile industry. This threat has prodded the industry into making the best of what it now has. A long-established, if sorry, fact is that American automakers will work for the public good only if they fear federal intervention or a successful competitor.

As Kenneth Hahn said in one of his letters to Arjay R. Miller, then president of Ford Motor Company, "It is my belief that you owe it to the citizens who have purchased your products to return to them in a positive manner some assistance in preserving their health. Any industry that creates an air pollution problem for the rest of the citizens should be the industry required to solve it."

And so the debate rages on all levels. One federal pollution control expert speaks of the many conferences on the subject he has attended. "They always remind me of the love life of an elephant," he says. "There's a lot of trumpeting at a very high level—but it takes a couple of years to see any tangible results."

Historically, the federal government has initiated (though usually belatedly) the most vigorous action against environmental freebooting; this has been the case in such areas as water pollution, pesticide poisoning, and soil erosion. But in the field of air pollution, one state—California—has taken the lead. The federal government simply has tagged along until now. Recently the leadership in the field has been grasped by the Department of Health, Education, and Welfare.

The most spirited battles in Congress have swirled around the concept of "federal standards." Standards are the limits set by government on the amount of pollutants which may be emitted by a particular source. In the past, such standards—with few exceptions—have been forcefully opposed by industry.

But the automakers recently have come around to the view that, if they must be saddled with emissions standards, a single set of federal regulations is preferable to dozens of different state laws with which they must comply. The Department of Health, Education, and Welfare established both exhaust and crankcase emission standards for all 1968 cars. The amount of exhaust emissions (which include hydrocarbons and carbon monoxide) permitted under these regulations will be further reduced by one-third on all 1970 cars (all pollutants from crankcases were to be eliminated in 1968 cars).

The application of federal standards to stationary sources of air pollution ran into trouble last year. The Department of Health, Education, and Welfare, speaking for the Johnson Administration, had sought such standards. But Maine Senator Edmund S. Muskie, chairman of the Senate Subcommittee on Air and Water Pollution (he is known as "Mr. Clean" in Washington), made one of his infrequent departures from the administration line.

Though he has fought for federal standards in water pollution legislation, Muskie retreated from them last year when the Air Quality Act was being pushed through Congress. He supported a plan to give *states* the first chance at establishing and enforcing clean air standards for entire "airsheds"—rather than post federal emission standards for individual polluters. He also opposed a proposal to impose fines of up to $1,000 a day on industry violators.

Muskie's retreat from strong national emission standards disappointed a number of pollution control experts in Washington, including some at Health, Education, and Welfare. But they were philosophical about it.

"It would be a mistake to criticize Muskie," an HEW official says. "He had help pushing *water* pollution control legislation through Congress. But air pollution is different. Many senators and representatives spouted industry's line. Muskie, more than anyone else in Congress, has stood up all the way for effective air pollution control."

The Air Quality Act that Congress passed in 1967 reflected Muskie's, rather than the Johnson Administration's, proposals. Federal emission standards were dropped. Nevertheless, the act gives the nation a comprehensive and rational approach to cleaning up its air, and the role of the states under this legislation is crucial. The country now must wait several years to see if the states have the courage to enforce their own standards. If not, the federal government will step in.

It is true, however, that as word of air pollution's contribution to chronic respiratory disease seeps through the public relations barrier, industry comes under more insistent pressures. Automobile manufacturers and the big oil companies have embarked on a number of research projects to bring their products more in line with what the crisis calls for. And one segment of industry, that which produces control devices for stationary sources (filters, wet scrubbers, electric precipitators for smokestacks), naturally is waiting anxiously to see the antipollution drive pressed home. Today these expanding companies do a business in excess of $150 million a year.

Citizen action groups, as the public is beginning to learn from past experiences, can exert enormous pressure on government as well as on industry. Conservation usually has been thought of as a movement limited to rural and suburban areas. But the concentration of deadly fumes in the cities has brought worried citizens together to fight for cleaner urban environments.

Organizations such as Stamp Out Smog (SOS) in Los Angeles, the Allegheny Conference on Community Development in Pittsburgh, and Citizens for Clean Air in New York City, often have served as the impetus for antipollution measures in their own cities. Concerned citizens are urged to contact their local action groups or write Clean Air, Washington, D.C.

Unfortunately, the leading conservation organizations have not yet spoken effectively in this area. Polluted water kills fish, pesticides kill birds and mammals—and so conservationists have kept to their traditional function of battling to preserve the natural world. Air pollution chiefly affects man. But it would be a tragic mistake to neglect air pollution control on that account. At the moment, man needs all the help he can get.

Citizens will act when they learn the seriousness of the problem. To date, neither the government, citizen action groups, nor conservation organizations have been able to get that story across. Sidney J. Edelman of the Department of Health, Education, and Welfare points as an example to a recent federal conference called in New York to abate air pollution. The government placed three ads in *The New York Times*, asking parties interested in making statements at the conference to notify the authorities of their intentions.

"But we only got one letter," Edelman said afterward. "That was from *The Times* billing us for the ads."

A number of the previous essays in this book have suggested that some substitute must be found for the burning of fossil fuels if massive and deadly air pollution buildups are to be avoided. One long-anticipated new source of power is atomic energy; indeed the prediction has been made that in thirty years half of our electric power may be provided by nuclear reactors, and the image of the friendly atom as our benign and trustworthy servant has been widely promoted by power companies and the Atomic Energy Commission. But in the face of these optimistic predictions, a growing number of scientists and critics are repeating what the authors of this article say: Nuclear power may pose the most grave environmental pollution threat yet. Several very serious shortcomings of nuclear power are set forth: for example, the possibility of catastrophic accidents, the dangers inherent in waste disposal, and the inevitable dispersion by nuclear plants of radiation into the surrounding air and waters. As George Woodwell also pointed out, radioactive isotopes may become concentrated as they move through a food chain causing dangerous buildups to occur even though only traces of radioactivity are being released. The authors call for investigation of new power sources rather than the proliferation of the poorly understood and potentially hazardous nuclear plants described here.

Richard Curtis (1937-   ) and Elizabeth Hogan are both free lance writers. Their detailed research for this essay included consultation with a number of nuclear scientists, engineers, and biologists.

THE MYTH OF THE PEACEFUL ATOM

by Richard Curtis and Elizabeth Hogan

"What is past is past, and the damage we may already have done to future generations cannot be rescinded, but we cannot shirk the compelling responsibility to determine if the course we are following is one we should be following."

So said Senator Thruston B. Morton of Kentucky on February 29, 1968, upon introducing into Congress a resolution calling for comprehensive review of federal participation in the atomic energy power program. Admitting he had been remiss in informing himself on this "grave danger," Morton said he had now looked more deeply into nuclear power safety and was "dismayed at some of the things I have found—warnings and facts from highly qualified people who firmly believe that we have moved too fast and without proper safeguards into an atomic power age."

Senator Morton's resolution on nuclear power was by no means the only one before Congress in 1968. Indeed, more than two dozen legislators urged investigation and re-evaluation of this program. This fact may come as a surprise to much of the public, for the belief is widespread that the nuclear reactors being built to generate electricity for our cities are safe,

---

"The Myth of the Peaceful Atom" From the March 1969 issue of *Natural History*. Reprinted by permission of the authors.

reliable, and pollution-free. But a rapidly growing number of physicists, biologists, engineers, public health officials, and even staff members of the Atomic Energy Commission itself—the government bureau responsible for regulation of this force—have been expressing serious misgivings about the planned proliferation of nuclear power plants. In fact, some have indicated that nuclear power, which Supreme Court Justices William O. Douglas and Hugo L. Black described as "the most deadly, the most dangerous process that man has ever conceived," represents the gravest pollution threat yet to our environment.

As of June, 1968, 15 commercial nuclear power plants were operating or operable within the United States, producing about one per cent of our current electrical output. The government, however, has been promoting a plan by which 25 per cent of our electric power will be generated by the atom by 1980, and half by the year 2000. To meet this goal, 87 more plants are under construction or on the drawing boards. Although atomic power and reactor technology are still imperfect sciences, saturated with hazards and unknowns, these reactors are going up in close proximity to heavy population concentrations. Most of them will be of a size never previously attempted by scientists and engineers. They are, in effect, gigantic nuclear experiments.

As most readers will recall, atomic reactors are designed to use the tremendous heat generated by splitting atoms. They are fueled with a concentrated form of uranium stored in thin-walled tubes bound together to form subassemblies. These are placed in the reactor's core, separated by control rods that absorb neutrons and thus help regulate chain reactions of splitting atoms. When the rods are withdrawn, the chain reactions intensify, producing enormous quantities of heat. Coolant circulated through the fuel elements in the reactor core carries the heat away to heat-exchange systems, where water is brought to a boil. The resultant steam is employed to turn electricity-generating turbines.

Stated in this condensed fashion, the process sounds innocuous enough. Unfortunately, however, heat is not the only form of energy produced by atomic fission. Another is radioactivity. During the course of operation, the fuel assemblies and other components in the reactor's core become intensely radioactive. Some of the fission by-products have been described as a million to a billion times more toxic than any known industrial chemical. Some 200 radioactive isotopes are produced as by-products of reactor operation, and the amount of just one of them, strontium-90, accumulated in a reactor of even modest (100-200 megawatt) size, after it has been operative for six months, is equal to what would be produced by the explosion of a bomb 190 times more powerful than the one dropped on Hiroshima.

Huge concentrations of radioactive material are also to be found in nuclear fuel-reprocessing plants. Because the intense radioactivity in a reactor core eventually interferes with the fuel's efficiency, the spent fuel assemblies must be removed from time to time and replaced by new, uncontaminated ones. The old ones are transported to reprocessing plants where the contaminants are separated from the salvageable fuel as well as from plutonium, a valuable by-product. Since no satisfactory means have been found for neutralizing or for safely releasing into the environment the radioactive liquid containing the contaminants, it must be stored until it is no longer dangerous. Thus, reprocessing plants and storage areas are immense repositories of "hot" and "dirty" material. Furthermore, routes between nuclear power plants and the reprocessing facility carry traffic bearing high quantities of such material.

Even from this glimpse it will be apparent that public and environmental safety depend on the flawless containment of radioactivity every step of the way. For, owing to the incredible potency of fission products, even the slightest leakage is harmful and a massive release would be catastrophic. The fundamental question, then, is how heavily can we rely on human wisdom, care, and engineering to hold this peril under absolute control?

Abundant evidence points to the conclusion that we cannot rely on it at all.

The hazards of peaceful atomic power fall into two broad categories: the threat of violent, massive releases of radioactivity or that of slow, but deadly, seepage of harmful products into the environment.

Nuclear physicists assure us that reactors cannot explode like atomic bombs because the complex apparatus for detonating an atomic warhead is absent. This fact, however, is of little consolation when it is realized that only a *conventional* explosion, which ruptures the reactor mechanism and its containment structure, could produce havoc on a scale eclipsing any industrial accident on record or any single act of war, including the atomic destruction of Hiroshima or Nagasaki.

There are numerous ways in which such an explosion can take place in a reactor. For example, liquid sodium, which is used in some reactors as a coolant, is a devilishly tricky element that under certain circumstances burns violently on contact with air. Accidental exposure of sodium could initiate a chain of reactions: rupturing fuel assemblies, damaging components and shielding, and destroying primary and secondary emergency safeguards. If coolant is lost, as it could be in some types of reactors, fuel could melt and recongeal, forming "puddles" that could explode upon reaching a critical size. If these explosions are forceful enough, and safeguards fail, some of the fission products could be released outside the plant and into the enviroment in the form of a gas or a cloud of fine radioactive particles. Under not uncommon atmospheric conditions such as an "inversion," in which a layer of warm air keeps a cooler layer from rising, a blanket of radioactivity could spread insidiously over the countryside. Another possibility is that fission products could be carried out of the reactor and into a city's watershed, for all reactors are being built on lakes, rivers, or other bodies of water for cooling purposes.

What would be the toll of such a calamity?

In 1957 the Atomic Energy Commission issued a study (designated Wash. —740), largely prepared by the Brookhaven National Laboratory, that attempted to assess the probabilities of such "incidents" and the potential consequences. Some of its findings were stupefying: From the explosion of a 100-200 megawatt reactor, as many as 3,400 people could be killed, 43,000 injured, and as much as 7 billion dollars of property damage done. People could be killed at distances up to 15 miles and injured up to 45. Land contamination could extend for far greater distances: agricultural quarantines might prevail over an area of 150,000 square miles, more than the combined areas of Pennsylvania, New York, and New Jersey.

The awful significance of these figures is difficult to comprehend. By way of comparison, we might look at one of the worst industrial accidents of modern times: the Texas City disaster of 1947 when a ship loaded with ammonium nitrate fertilizer exploded, virtually leveling the city, killing 561 people, and causing an estimated $67 million worth of damage. Appalling as this catastrophe was, however, it does not begin to approach the potential havoc that would be wreaked by a nuclear explosion occurring in one of the plants now being constructed close to several American cities.

The scientists and engineers who produced the Brookhaven Report optimistically ventured to give high odds against such an occurrence, asserting that the structures, systems, and safeguards of atomic plants were so engineered as to render it practically incredible. At the same time, though, the report was replete with such statements as:

"The cumulative effect of radiation on physical and chemical properties of materials, after long periods of time, is largely unknown."

"Much remains to be learned about the characteristics and behavior of nuclear systems."

"It is important to recognize that the magnitudes of many of the crucial factors in this study are not quantitatively established, either by theoretical and experimental data or adequate experience."

Even if the report had been founded on more substantial understanding of natural and technical processes, many of the grounds on which the Brookhaven team based its conclusions are shaky at best.

For one thing, all of us are familiar with technological disasters that have occurred against fantastically high odds: the sinking of the "unsinkable" *Titanic*, or the November 9, 1965, "blackout" of the northeastern United States, for example. The latter happening illustrates how an "incredible" event can occur in the electric utility field, most experts agreeing that the chain of circumstances that brought it about was so improbable that the odds against it defy calculation.

Congressional testimony given in 1967 by Dr. David Okrent, a former chairman of the AEC's Advisory Committee on Reactor Safeguards, demonstrated that fate is not always a respecter of enormously adverse odds. "We do have on record cases where, for example, an applicant, appearing before an atomic safety and licensing board, stated that a mathematical impossibility had occurred; namely, one tornado took out five separate power lines to a reactor. If one calculated strictly on the basis of probability and multiplied the probability for one line five times, you get a very small number indeed," said Dr. Okrent, "but it happened."

A disturbing number of reactor accidents have occurred—with sheer luck playing an important part in averting catastrophe—that seem to have been the product of incredible coincidences. On October 10, 1957, for instance, the Number One Pile (reactor) at the Windscale Works in England malfunctioned, spewing fission products over so much territory that authorities had to seize all milk and growing foodstuffs in a 400-square-mile area around the plant. A British report on the incident stated that *all* of the reactor's containment features had failed. And, closer to home, a meltdown of fuel in the Fermi reactor in Lagoona Beach, Michigan, in October, 1966, came within an ace of turning into a nuclear "runaway." An explosive release of radioactive materials was averted, but the failures of Fermi's safeguards made the event, in the words of Sheldon Novick in *Scientist and Citizen*, "a bit worse than the 'maximum credible accident.'"

The atomic industry has attempted to design components and safeguards so that failure of one vital system in a plant will not affect another, resulting in a "house of cards" collapse. However, two highly regarded authorities, Theos J. Thompson and J. G. Beckerley, in a book on reactor safety advise us not to place too much faith in claims of independent components." Many manufacturers and utility operators have resisted the idea of producing "redundant safeguards" on the grounds of excessive cost.

Investigations of reactor breakdowns usually disclose a number of small, seemingly unrelated failures, which snowballed into one big one. A design flaw or a human error, a component failure here, an instrumentation failure there—all may coincide to contribute to the total event. Thompson and Beckerley, examining several atomic plant accidents, pinpointed 13 different contributing causes in three of the accidents that had occurred up to the time of their 1964 study.

Among the many factors contributing to reactor accidents, the human element is the most difficult to quantify. And perhaps for that reason, it has been largely overlooked in the AEC's assessments of reactor safety. Yet, a private researcher of nuclear accidents, Dr. Donald Oken, M.D., Associate Director of the Psychosomatic and Psychiatric Institute of Michael Reese Hospital in Chicago reported: "A review of reports of past criticality and reactor incidents and discussions held with some of the health personnel in charge reveal a number of striking peculiarities in the behavior of many of those involved—in which they almost literally asked for trouble."

AEC annuals are full of reports of human negligence: 3,844 pounds of uranium hexafluoride lost owing to an error in opening a cylinder; a $220,000 fire in a reactor because of accidental tripping of valves by electricians during previous maintenance work; numerous vehicular accidents involving transport of nuclear materials. None of these accidents led to disaster, but who will warrant that, with the projected proliferation of power plants and satellite industries in the coming decade, a moment's misjudgment will not trigger a nightmare? Perhaps worse, the likelihood of sabotage has scarcely been weighed, despite a number of incidents and threats.

It should be apparent that if men are to build safe, successful reactors, the whole level of industrial workmanship, engineering, inspection, and quality control must be raised well above prevailing levels. The more sophisticated the technology, the more precise the correspondence between the subtlest gradations of care or negligence and that technology's success or failure. When meters, grams, and seconds are no longer good enough, and specifications call for millimeters, milligrams, and milliseconds, the demands made on men, material, and machinery are accordingly intensified. Minute lapses that might be tolerable in a conventional industrial procedure will wreck the more exacting one. And when the technology is not only exacting but hazardous in the extreme, then a trivial oversight, a minor defect, a moment's inattention may spell doom.

While there is little doubt that American technology is the most refined on earth, there is ample reason to believe that it has more than met its match in the seemingly insurmountable problems posed by the peaceful atom. Societies of professional engineers, and others concerned with establishing technical and safety criteria for the nuclear industry, have described between 2,800 and 5,000 technical standards that are necessary for a typical reactor power plant in such areas as materials, testing, design, electrical gear, instrumentation, plant equipment, and processes. Yet, due to the rapidity with which the nuclear industry has developed, as of March, 1967, only about 100 of these had been passed on and approved for use.

It is not surprising, then, to learn that serious technical difficulties are turning up in reactor after reactor. At the Big Rock Point Nuclear Plant, a relatively small reactor near Charlevoix, Michigan, control rods were found sticking in position, studs failing or cracked, screws jostled out of place and into key mechanisms, a valve malfunctioning for more than a dozen reasons, foreign material lodging in critical moving parts, and welds cracked on every one of sixteen screws holding two components in place. A reactor at Humboldt Bay in California manifested cracks in the tubes containing fuel: in order to keep costs down, stainless steel had been used instead of a more reliable alloy. The Oyster Creek plant in New Jersey showed cracks in 123 of 137 fuel tubes, and welding defects at every point where tubes and control-rod housings were joined around the reactor's vessel. Reactors in Wisconsin, Minnesota, Connecticut, Puerto Rico, New York, and elsewhere have experienced innumerable operating difficulties, and some, such as the

$55 million Hallam plant in Nebraska have been forced to shut down for good, owing to plant malfunction.

Chilling parallels can be drawn between failures in nuclear utility technology and in the nuclear submarine program. In October, 1962, Vice Admiral Hyman G. Rickover, Director of AEC's Division of Naval Reactors, took the atomic industry to task in a speech in New York City:

"It is not well enough understood that conventional components of advanced systems must necessarily meet higher standards. Yet it should be obvious that failures that would be trivial if they occurred in a conventional application will have serious consequences in a nuclear plant because here radioactivity is involved. . . ."

Rickover went on to cite defective welds, forging materials substituted without authorization, violations of official specifications, poor inspection techniques, small and seemingly "unimportant" parts left out of components, faulty brazing of wires, and more. "I assure you," he declared, "I am not exaggerating the situation; in fact, I have understated it. For every case I have given, I could cite a dozen more."

The following April, the U.S. atomic submarine *Thresher*, while undergoing a deep test dive some 200 miles off the Cape Cod coast, went down with 112 naval personnel and 17 civilians and never came up again. Subsequent investigation revealed that the sub suffered from many of the same ailments described in Rickover's speech. "It is extremely unfortunate," said Senator John O. Pastore, chairman of the joint congressional committee that held hearings on the disaster, "that this tragedy had to occur to bring a number of unsatisfactory conditions into the open." We must now ask if the same will one day be said about a power plant near one of our large cities.

If a major reactor catastrophe did occur there is good reason to believe that the consequences would be far worse than even the dismaying toll suggested by the 1957 Brookhaven Report, for a number of developments since then have made the threat considerably more formidable.

The Brookhaven Report's accident statistics, for instance, pertained to a reactor of between 100 and 200 megawatts. But while the 15 reactors currently operating in the United States average about 186 megawatts, the 87 plants going up or planned for the next decade are many times that size. Thirty-one under construction average about 726 megawatts; 42 in the planning stage average 832; 14 more, planned but without reactors ordered, will average 904. Some, such as those slated for Illinois, California, Alabama, and New York anticipate capacities of more than 1,000 megawatts. Con Edison has just announced it intends to build four units of 1,000 megawatts each on Long Island Sound near New Rochelle in teeming Westchester County—four nuclear reactors, each with a capacity five to ten times that of the reactor described in the Brookhaven Report.

These facilities will accordingly contain more uranium fuel, and because it is costly to replace spent fuel assemblies (this delicate and dangerous process can take six weeks or longer), the new reactors are designed to operate without fuel replacement far beyond the six months posited in the Brookhaven Report. As a result, the buildup of toxic fission products in tomorrow's reactors will be far greater than at present, and an accident occurring close to the end of the "fuel cycle" in such a plant could release fantastic amounts of radioactive material.

Most serious of all, perhaps, is that tomorrow's reactors are now slated for location in close proximity to population concentrations. While the Brookhaven Report had its hypothetical reactor situated about 30 miles from a major city, many of tomorrow's atomic plants will be much closer. Although the AEC has drafted "guidelines" for siting reactors, the Commission has failed to make utilities adhere to them. In 1967, Clifford K. Beck, AEC's Deputy Director of Regulation, admitted to the Joint Committee on Atomic Energy that nuclear plants in Connecticut, California, New York, and other locations "have been approved with lower distances than our general guides would have indicated when they were approved."

Also, we must remember that while a reactor may not be near the legal boundaries of a metropolis, it may lie close to a population center. Thus, while Con Edison's Indian Point plant is 24 miles from New York City (two more plants are now being built there), it is within 10 miles of an estimated population of 155,510. It need only be recalled that the Brookhaven Report foresaw people being killed by a major radioactive release at distances up to 15 miles to realize the significance of these figures.

In a recent study of nuclear plant siting made by W. K. Davis and J. E. Robb of San Francisco's Bechtel Corporation, the locations of 42 nuclear power plants (some proposed, some now operable) were examined with respect to population centers inhabited by 25,000 residents or more. Their findings are unnerving: only *two* plants in operation or planned are more than 30 miles from a population center. Of the rest, 14 are between 20 and 27 miles away, 15 between 10 and 16 miles, and 11 between 1 and 9 miles.

Is it necessary to build atomic plants so big and so close? The answer has to do with economics. The larger a facility is, the lower the unit cost of construction and operation and the cheaper the electricity. The longer the fuel cycle, the fewer the expensive shutdowns while spent fuel assemblies are replaced. The closer the plant is to the consumer, the lower the cost of rights of way, power lines, and other transmission equipment.

On a few occasions an aroused public has successfully opposed the situation of plants near population centers. When the Pacific Gas and Electric Company persisted in trying to build a reactor squarely over earthquake faults in an area of known seismic activity—the site was Bodega Head, north of San Francisco—a courageous conservation group forced the company to back down. It has been suggested, though, that the group might not have won had not the Alaskan earthquake of 1964, occurring while the fight was going on, underscored the recklessness of the utility's scheme.

Announcement by Con Edison at the end of 1962 of its proposal to build a large nuclear plant in Ravenswood, Queens, close to the center of New York City brought a storm of frightened and angry protest. Although the utility's chairman noted, "We are confident that a nuclear plant can be built in Long Island City, or in Times Square for that matter, without hazard to our own employees working in the plant or to the community," David E. Lilienthal, the former head of the AEC, had a contrary opinion, declaring he "would not dream of living in Queens if a huge nuclear plant were located there." Outraged citizens and a number of noted scientists prevailed.

For the most part, however, the battle has been a losing one. Con Edison, for example, after its defeat in the Ravenswood fight, has just announced an interest in building a reactor on Welfare Island, literally a stone's throw from midtown Manhattan. Also, New York's Governor Nelson Rockefeller has gone on record advocating an $8 billion electric power expansion program based extensively on nuclear energy. The state legislature approved of the program, and in 1968, voted to bolster the plan with state subsidies.

Some of the deepest concern about the size and location of atomic plants has been expressed by members of the AEC themselves. "The actual experience with reactors in general is still quite limited," said Harold Price, AEC's Director of Regulation, in 1967 congressional hearings, "and with large reactors of the type now being considered, it is non-existent. Therefore, because there would be a large number of people close by and because of lack of experience, it is . . . a matter of judgment and prudence at present to locate reactors where the protection of distance will be present."

Price's statement is mild compared to that made in the same hearings by Nunzio J. Palladino, Chairman of the AEC's Advisory Committee on Reactor Safeguards for 1967, and Dr. David Okrent, former Chairman for 1966: "the ACRS believes that placing large nuclear reactors close to population centers will require considerable further improvements in safety, and that *none of the large power reactors now under construction is considered suitable for location in metropolitan areas* [our italics]."

The threat of a nuclear plant catastrophe constitutes only half of the double jeopardy in which atomic power has placed us. For even if no such calamity occurs, the gradual exhaustion of what one scientist terms our environmental "radiation budget," due to unavoidable releases of radioactivity during normal operation of nuclear facilities, poses an equal and possibly more insidious threat to all living things on earth.

Most of the fission products created in a reactor are trapped. Contaminated solids, liquids, and gasses are isolated, allowed to decay for a short period of time, then concentrated and shipped in drums to storage areas. These are called "high-level wastes." But technology for retaining all radioactive contaminants, is either unperfected or costly, and much material of low-level radioactivity is routinely released into the air or water at the reactor site. These releases are undertaken in such a way, we are told, as to insure dispersion or dilution sufficient to prevent any predictable human exposure above harmful levels. Thus, when atomic power advocates are asked about the dangers of contaminating the environment, they imply that the relatively small amounts of radioactive materials released under "planned" conditions are harmless.

This view is a myth.

In the first place, many waste radionuclides take an extraordinarily long time to decay. The half-life (the time it takes for half of an element's atoms to disintegrate through fission) of strontium-90, for instance, is more than 27 years. Thus, even though certain long-lived isotopes are widely dispersed in air or diluted in water, their radioactivity does not cease. It remains, and over a period of time accumulates. It is therefore not pertinent to talk about the safety of any single release of "hot" effluents into the environment. At issue, rather, is their duration and cumulative radioactivity.

Further, many radioactive elements taken into the body tend to build up in specific tissues and organs to which those isotopes are attracted, increasing by many times the exposure dosage in those local areas of the body. Iodine-131, for instance, seeks the thyroid gland; strontium-90 collects in the bones; cesium-137 accumulates in muscle. Many isotopes have long half-lives, some measurable in decades.

Two more factors controvert the view that carefully monitored releases of low-level radioactivity into the environment are not pernicious. First, there is apparently no radiation threshold below which harm is impossible. Any dose, however small, will take its toll of cell material, and that damage is irreversible. Second, it may take decades for organic damage, or generations for genetic damage, to manifest itself. In 1955, for example, two British doctors reported a case of skin cancer—ultimately fatal—that had taken forty-nine years to develop following fluoroscopic irradiation of a patient.

Still another problem has received inadequate attention. Man is by no means the only creature in whom radioactive isotopes concentrate. The dietary needs of all plant and animal life dictate intake of specific elements. These concentrate even in the lowest and most basic forms of life. They are then passed up food chains, from grass to cattle to milk to man, for example. As they progress up these chains, the concentrations often increase, sometimes by hundreds of thousands of times. And if these elements are radioactive. . . .

Take zinc-65, produced in a reactor when atomic particles interact with zinc in certain components. Scrutiny of the wildlife in a pond receiving runoff from the Savannah River Plant near Aiken, South Carolina, disclosed that while the water in that pond contained only infinitesimal traces of radioactive zinc-65, the algae that lived on the water had concentrated the isotope by nearly 6,000 times. The bones of bluegills, an omnivorous fish that feeds both on algae and on algae-eating fish, showed concentrations more than 8,200 times higher than the amount found in the water. Study of the Columbia River, on which the Hanford, Washington, reactor is located, revealed that while the radioactivity of the water was relatively insignificant: 1. the radioactivity of the river plankton was 2,000 times greater; 2. the radioactivity of the fish and ducks feeding on the plankton was 15,000 and 40,000 times greater, respectively; 3. the radioactivity of young swallows fed on insects caught by their parents near the river was 500,000 times greater; and 4. the radioactivity of the egg yolks of water birds was more than a million times greater.

Here then are clear illustrations of the ways in which almost undetectable traces of radioactivity in air, water, or soil may be progressively concentrated, so that by the time it ends up on man's plate or in his glass it is a tidy package of poison.

That nuclear facilities are producing dangerous buildups of radio-isotopes in our environment can be amply documented. University of Nevada investigators, seeking a cause for concentrations of iodine-131 in cattle thyroids in wide areas of the western United States, concluded that "the principal known source of I-131 that could contribute to this level is exhaust gases from nuclear reactors and associated fuel-processing plants."

In his keynote address to the Health Physics Society Symposium at Atlanta, Georgia, early in 1968, AEC Commissioner Wilfred E. Johnson admitted that the release into the atmosphere of tritium and noble gases such as krypton-85 would present a potential problem in the future, and that, as yet, scientists had not devised a way of solving it. Krypton-85, although inert, has a 10-year half-life and tends to dissolve in fatty tissue, meaning fairly even distribution throughout the human body. Krypton-85 is particularly difficult to filter out of reactor discharges, and the accumulation of this element alone may exhaust as much as two-thirds of the "average" human's "radiation budget" for the coming century, based on the standards established by the National Committee on Radiation Protection and Measurement.

That "low-level" waste is a grossly deceptive term is obvious. In his book *Living with the Atom*, author Ritchie Calder in 1962 described an "audit" of environmental radiation that he and his colleagues, meeting at a symposium in Chicago, drew up to assess then current and future amounts of radioactivity released into atmosphere and water. Speculations covered the period 1955-65, and because atomic power plants were few and small during that time, the figures are more significant in relation to the future. Tallying "planned releases" of radiation from such sources as commercial and test reactors, nuclear ships, uranium mills, plutonium factories, and fuel-reprocessing plants, Calder's group came to a most disquieting conclusion: "By the time we had added up all the curies which might predictably be released, by all those peaceful uses, into the environment, it came to about 13 million curies per annum." A "curie" is a standard unit of radioactivity whose lethality can be appreciated from the fact that one trillionth of one curie of radioactive gas per cubic meter of air in a uranium mine is ten times higher than the official maximum permissible dose.

Calder's figures did not include fallout due to bomb testing and similar experiments, nor did they take into account possible reactor or nuclear transportation accidents. Above all, they did not include possible escape of stored high-level radioactive wastes, the implications of which were awesome to contemplate: "what kept nagging us was the question of waste disposal and of the remaining radioactivity which must not get loose. We were told that the dangerous waste, which is kept in storage, amounted to 10,000 million curies. If you wanted to play 'the numbers game' as an irresponsible exercise, you could divide this by the population of the world and find that it is over 3 curies for every individual."

Exactly what does Calder mean by "the question of waste disposal"?

It has been estimated that a ton of spent fuel in reprocessing will produce from forty to several hundred gallons of waste. This substance is a violently lethal mixture of short- and long-lived isotopes. It would take five cubic miles of water to dilute the waste from just *one* ton of fuel to a safe concentration. Or, if we permitted it to decay naturally until it reached the safe level—and the word "safe" is used advisedly—just one of the isotopes, strontium-90, would still be damaging to life 1,000 years from now, when it will have only one seventeen-billionth of its current potency.

There is no known way to reduce the toxicity of these isotopes; they must decay naturally, meaning *virtually perpetual containment*. Unfortunately, mankind has exhibited little skill in perpetual creations, and procedures for handling radioactive wastes leave everything to be desired. Formerly dumped in the ocean, the most common practice today is to store the concentrates in large steel tanks shielded by earth and concrete. This method has been employed for some twenty years, and about 80 million gallons of waste are now in storage in about 200 tanks. This "liquor" generates so much heat it boils by itself for years. Most of the inventory in these caldrons is waste from weapons production, but within thirty years, the accumulation from commercial nuclear power will soar if we embark upon the expansion program now being promoted by the AEC. Dr. Donald R. Chadwick, chief of the Division of Health of the U.S. Public Health Service, estimated in 1963 that the accumulated volume of waste material would come to two billion gallons by 1995.

It is not just the volume that fills one with sickening apprehension but the techniques of disposing of this material. David Lilienthal put his finger on the crux of the matter when he stated: "These huge quantities of radioactive wastes must somehow be removed from the reactors, must—

without mishap—be put into containers that will never rupture; then these vast quantities of poisonous stuff must be moved either to a burial ground or to reprocessing and concentration plants, handled again, and disposed of, by burial or otherwise, with a risk of human error at every step." Nor can it be stressed strongly enough that we are not discussing a brief danger period of days, months, or years. We are talking of periods "longer," in the words of AEC Commissioner Wilfred E. Johnson, "than the history of most governments that the world has seen."

Yet already there are many instances of the failure of storage facilities. An article in an AEC publication has cited nine cases of tank failure out of 183 tanks located in Washington, South Carolina, and Idaho. And a passage in the AEC's authorizing legislation for 1968 called for funding of $2,500,000 for the replacement of failed and failing tanks in Richland, Washington. "There is no assurance," concluded the passage, "that the need for new waste storage tanks can be forestalled." If this is the case after twenty years of storage experience, it is beyond belief that this burden will be borne without some storage failures for centuries in the future. Remember too, that these waste-holding "tank farms" are vulnerable to natural catastrophes such as earthquakes, and to man-made ones such as sabotage.

Efforts are of course being made toward effective handling of the waste problems, but many technical barriers must still be overcome. It is unlikely they will all be overcome by the end of the century, when waste tanks will boil with 6 billion curies of strontium-90, 5.3 billion curies of cesium-137, 6.07 billion curies of prometheum-147, 10.1 billion curies of cerium-144, and millions of curies of other isotopes. The amount of strontium-90 alone is 30 times more than would be released by the nuclear war envisioned in a 1959 congressional hearing.

The burden that radioactive wastes place on future generations is cruel and may prove intolerable. Physicist Joel A. Snow stated it well when he wrote in *Scientist and Citizen:* "Over periods of hundreds of years it is impossible to ensure that society will remain responsive to the problems created by the legacy of nuclear waste which we have left behind."

"Legacy" is indeed a gracious way of describing the reality of this situation, for at the very least we are saddling our children and their descendants with perpetual custodianship of our atomic refuse, and at worst may be dooming them to the same agonizing afflictions and deaths suffered by those who survived Hiroshima. Radiation has been positively linked to cancer, leukemia, brain damage, infant mortality, cataracts, sterility, genetic defects and mutations, and general shortening of life.

The implications for the survival of mankind can be glimpsed by considering just one of these effects, the genetic. In a 1960 article, James F. Crow, Professor of Genetics at the University of Wisconsin School of Medicine and president of the Genetics Society of America, stated that for every roentgen of slow radiation—the kind we can expect to receive in increasing doses from peacetime nuclear activity—about five mutations will result per 100 million genes exposed, meaning that "after a number of generations of exposure to one roentgen per generation, about one in 8,000 . . . in each generation would have severe genetic defects attributable to the radiation."

The Atomic Energy Commission is aware of the many objections that have been raised to the atomic power program: why does it continue to encourage it? Unfortunately, the Commission must perform two conflicting roles. On the one hand, it is responsible for regulating the atomic power industry. But on the other, it has been charged by Congress to promote the use of nuclear energy by the utility industry. Because of its involvement in the highest priorities of national security, enormous power and legislative advantages have been vested in the AEC, enabling it to fulfill its role as promoter with almost unhampered success—while its effectiveness as regulator has gradually atrophied. The Commission consistently denies claims that atomic power is heading for troubled waters, optimistically reassuring critics that these plants are safe, clean neighbors.

The fact that there is no foundation for this optimism is emphasized by the insurance situation on atomic facilities. Despite the AEC's own assertion that as much as $7 billion in property damage could result from an atomic power plant catastrophe, the insurance industry, working through two pools, will put up no more than $74 million, or about one per cent, to indemnify equipment manufacturers and utility operators against damage suits from the public. The federal government will add up to $486 million more, but this still leaves more than $6 billion in property damages to be picked up by victims of a Brookhaven-sized accident. And no insurance company—not even Lloyds of London—will issue property insurance to individuals against radiation damage. If there is so little risk in atomic power plants, why is insurance so inadequate?

The knowledge that man must henceforth live in constant dread of a major nuclear plant accident is disturbing enough. But we must recognize that even if such calamities are averted, the slow saturation of our environment with radioactive wastes will nevertheless be raising the odds that you or your heirs will fall victim to one of a multitude of afflictions. There is no "threshold" exposure below which we can feel safe.

We have little time to reflect on our alternatives, for the moment must soon come when no reversal will be possible. Dr. L. P. Hatch of Brookhaven National Laboratory vividly made this point when he told the Joint Committee on Atomic Energy: "If we were to go on for 50 years in the atomic power industry, and find that we had reached an impasse, that we had been doing the wrong thing with the wastes and we would like to reconsider the disposal methods, it would be entirely too late, because the problem would exist and nothing could be done to change that fact for the next, say, 600 or a thousand years." To which might be added a sobering thought stated by Dr. David Price of the U.S. Public Health Service: "We all live under the haunting fear that something may corrupt the environment to the point where man joins the dinosaurs as an obsolete form of life. And what makes these thoughts all the more disturbing is the knowledge that our fate could perhaps be sealed twenty or more years before the development of symptoms."

What must be done to avert the perils of the peaceful atom? A number of plans have been put forward for stricter regulation of activities in the nuclear utility field, such as limiting the size of reactors or their proximity to population concentrations or building more safeguards. As sensible as these proposals appear on the surface, they fail to recognize a number of important realities: first, that such arrangements would probably be opposed by utility operators and the government due to their prohibitively high costs. Since our government seems to be committed to making atomic power plants competitive with conventionally fueled plants, and because businesses are in business for profit, it is hardly likely they would buy these answers. Second, the technical problems involved in containment of radioactivity have not been successfully overcome, and there is little likelihood they will be resolved in time to prevent immense and irrevocable harm to our environment. Third, the nature of business enterprise is unfortunately such that *perfect* policing of the atomic power industry is unachievable. As we have seen in the cases of other forms of pollution, the public spirit of men seeking profit from industrial processes does not always rise as high as the welfare of society requires. It is unwise to hope that stricter regulation would do the job.

What, then, is the answer? The only course may be to turn boldly away from atomic energy as a major source of electricity production, abandoning it as this nation has abandoned other costly but unsuccessful technological enterprises.

There is no doubt that, with this nation's demand for electricity doubling every decade, new power sources are urgently needed. Nor is there doubt that our conventional fuel reserves—coal, oil, and natural gas—are rapidly being consumed. Sufficient high-grade fossil fuel reserves exist, however, to carry us to the end of this century; and new techniques for recovering these fuels from secondary sources such as oil shale could extend the time even longer. Furthermore, advances in pollution abatement technology and revolutionary new techniques, now in development, for burning conventional fuels with high efficiency, could carry us well into the next century with the fossil fuels we have. This abundance, and potential abundance, gives us at least several decades to survey possible alternatives to atomic power, select the most promising, and develop them on an appropriate scale as alternatives to nuclear power. Solar energy, tidal power, heat from the earth's core, and even garbage and solid-waste incineration have to some degree been demonstrated as promising means of electricity generation. If we subsidized research and development of those fields as liberally as we have done atomic energy, some of them would undoubtedly prove to be what atomic energy once promised, without its deadly drawbacks.

Aside from the positive prospect of profitability in these new approaches, industry will have another powerful incentive for turning to them; namely, that atomic energy is proving to be quite the opposite of the cheap, everlasting resource envisioned at the outset of the atomic age. The prices of reactors and components and costs of construction and operation have soared in the last few years, greatly damaging nuclear power's position as a competitor with conventional fuels. If insurance premiums and other indirect subsidies are brought into line with realistic estimates of what it takes to make atomic energy both safe and economical, the atom might prove to be the most *expensive* form of energy yet devised—not the cheapest. In addition, because of our wasteful fuel policies, evidence indicates that sources of low-cost uranium will be exhausted before the turn of century. Fuel-producing breeder reactors, in which the nuclear establishment has invested such high hopes for the creation of vast, new fuel supplies, have proved a distinct technological disappointment. Even if the problems plaguing this effort were overcome in the next ten or twenty years, it may still be too late to recoup the losses of nuclear fuel reserves brought about by prodigious mismanagement.

The proposal to abandon or severely curtail the use of atomic energy is clearly a difficult one to imagine. We have only to realize, however, that by pursuing our current civilian nuclear power program, we are jeopardizing every other industry in the country; in that light, this proposal becomes the only practical alternative. In short, the entire national community stands to benefit from the abandonment of a policy which seems to be leading us toward both environmental and economic disaster.

Man's incomplete understanding of many technological principles and natural forces is not necessarily to his discredit. Indeed, that he has erected empires despite his limited knowledge is to his glory. But that he pits this ignorance and uncertainty, and the fragile yet lethal technology he has woven out of them, against the uncertainties of nature, science, and human behavior—this may well be to his everlasting sorrow.

"In Wildness
is the Preservation
of the World"

Conservationists and other defenders of what Henry David Thoreau called "Wildness" are often dismissed by their detractors with epithets such as "nature-lover" which suggest that anyone who does seek and study nature must be considered both irrelevant and befuddled. Behind this attitude is the view that nature itself is merely pretty and pleasant. Joseph Wood Krutch (1893-1970) contradicts both of these assumptions: First, nature, as Thoreau rightly understood it, is a titanic and elemental force; second, the understanding of nature is the most relevant basis for rational human behavior. Krutch reminds the reader that nature's plan did work fairly well, but it is not certain that man's plan will; indeed the evidence presented by the essays in this book is that man's plan is not working well at all. Of course, nature may yet have the final word: Recall Paul Ehrlich's closing caveat in "Eco-Catastrophe!" that "Nature bats last!" Krutch's essay and the three others which comprise this section argue for a wider recognition of man's dependence upon nature, his involvement in wildness.

Joseph Wood Krutch achieved considerable reputation as a literary scholar and intellectual historian in his earlier years. His later interests led him into the field of natural science, in which he has published widely.

WILDERNESS AS A TONIC

by Joseph Wood Krutch

A few months ago the Sierra Club of California published a magnificent book of seventy-two large color photographs by Eliot Porter. Though I wrote a brief introduction for it, this is not a plug. As a matter of fact, my subject did not begin to worry me until the introduction was in type and I learned for the first time what the title of the book was to be: *In Wildness Is the Preservation of the World*.

"Oh, dear," I said to myself, "that's a foolish title." Nature-lover though I am, this was going too far. "Wildness," I said, "is a tonic and a refreshment. I think we are losing something as it disappears from our environment. But to call it 'the preservation of the world' is pretty far-fetched. True, I doubt that we can be saved by increased production or by trips to the moon. In fact, I sometimes doubt that we can be saved at all. But wildness! Now really!"

Another shock came when I learned that the phrase is a quotation from Henry David Thoreau. Henry, I know, confessed his love of exaggeration and once said his only fear was the fear he might not be extravagant

---

enough. But he rarely said anything really foolish. I thought I had better look up the context. And not to keep the reader in suspense (if I may flatter myself that I have generated any), I must confess that I have come to the conclusion that what Henry said is neither foolish nor exaggerated. It is a truth almost as obvious as "the mass of men lead lives of quiet desperation."

Here is the context:

> *The West of which I speak is but another name for the Wild, and what I have been preparing to say is, that in Wildness is the preservation of the world. Every tree sends its fiber forth in search of the Wild. The cities import it at any price. Men plow and sail for it. From the forests and wilderness come the tonics and barks which brace mankind.*

What is this wildness Thoreau is talking about? It is not D. H. Lawrence's Dark Gods and neither is it the mindless anarchy of some current anti-intellectuals. Those are destructive forces. Thoreau's wildness is, on the other hand, something more nearly akin to Bernard Shaw's Life Force—it is that something prehuman that generated humanity. From it came a magnificent complex of living things long, long before we were here to be aware of them, and still longer before we, in our arrogance, began to boast that we were now ready to take over completely; that henceforth we in our greater wisdom would plan and manage everything, even, as we sometimes say, direct the course of evolution itself.

Yet if—as seems not unlikely—we should manage or mismanage to the point of self-destruction, wildness alone will survive to make a new world.

Like so much that Thoreau wrote, this appeal to wildness as the ultimate hope for survival is more relevant and comprehensible in the context of our world than it was in his. A few of his contemporaries—Melville and Hawthorne, for instance—may have already begun to question the success of the human enterprise and the inevitability of progress. But life as it is managed by man was becoming increasingly comfortable and seemingly secure. Men were often anxious (or, as Thoreau said, "desperate") concerning their individual lives. But few doubted that mankind *as a whole* was on the right track. Theirs was not yet an age of public, overall anxiety. That mankind might plan itself into suicide occurred to nobody. Yet, though few today put such hopes as they have managed to maintain in anything except the completer dependence upon human institutions and inventions, the possibility that we have too little faith in "wildness" is not quite so preposterous a suspicion as it seemed then.

In a world that sits not on a powder keg but on a hydrogen bomb, one begins to suspect that the technician who rules our world is not the master magician he thinks he is but only a sorcerer's apprentice who does not know how to turn off what he turned on—or even how to avoid blowing himself up.

Should that be what he at last succeeds in doing, it would be a relatively small disaster compared with the possibility that he might destroy at the same time all that "wildness" that generated him and might in time generate something better. Perhaps there is life (or shall we call it "wildness"?) on other planets, but I hope it will remain ours, too. "Pile up your books, the records of sadness, your saws and your laws. Nature is glad outside, and her merry worms will ere long topple them down."

This wildness may often be red in tooth and claw. It may be shockingly careful of the type but careless of the single life. In this and in many other respects we are unwilling to submit to it. But somehow it did, in the end, create the very creatures who now criticize and reject it. And it is not so certain as it once seemed that we can successfully substitute entirely our competence for nature's. Nature is, after all, the great reservoir of energy, of confidence, of endless hope, and of that joy not wholly subdued by the pale cast of thought that seems to be disappearing from our human world. Rough and brutal though she sometimes seems in her far from simple plan, it did work, and it is not certain that our own plans will.

A very popular concept today is embodied in the magic word cybernetic—or self-regulating. "Feedback" is the secret of our most astonishing machines. But the famous balance of nature is the most extraordinary of all cybernetic systems. Left to itself, it is always self-regulated. The society we have created is not, on the other hand, cybernetic at all. The wisest and most benevolent of our plannings requires constant attention. We must pass this or that law or regulation, we must redress this balance of production and distribution, taking care that encouraging one thing does not discourage something else. The society we have created puts us in constant danger lest we ultimately find ourselves unable to direct the more and more complicated apparatus we have devised.

A really healthy society, so Thoreau once wrote, would be like a healthy body that functions perfectly without our being aware of it. We, on the other hand, are coming more and more to assume that the healthiest society is one in which all citizens devote so much of their time to arguing, weighing, investigating, voting, propagandizing, and signing protests in a constant effort to keep a valetudinarian body politic functioning in some sort of psuedo-health that they have none of that margin for mere living that Thoreau thought important. It's no wonder that such a situation generated beatniks by way of a reaction.

Many will no doubt reply that Thoreau's ideal sounds too close to that of the classical economists who trusted the cybernetic free competition of Herbert Spencer *et al.* and that it just doesn't work. But is it certain that our own contrary system is working very well when it produces, on the one hand, a more or less successful welfare state and, on the other, an international situation that threatens not only welfare but human existence itself?

Should the human being turn out to be the failure some began a generation or more ago to call him, then all is not necessarily lost. Unless life itself is extinguished, nature may begin where she began so long ago and struggle upward again. When we dream of a possible superman we almost invariably think of him as a direct descendant of ourselves. But he might be the flower and the fruit of some branch of the tree of life now represented only by one of the "lower" (and not necessarily anthropoid) animals.

When I recommend that we have a little more faith in the ultimate wisdom of nature I am not suggesting that national parks, camping trips, and better bird-watching are the best hope of mankind. But I do believe them useful reminders that we did not make the world we live in and that its beauty and joy, as well as its enormous potentialities, do not depend entirely on us. "Communion with nature" is not merely an empty phrase. It is the best corrective for that *hubris* from which the race of men increasingly suffers. Gerard Manley Hopkins, unwavering Catholic Christian though he was, could write:

> *What would the world be, once bereft*
> *Of wet and wildness? Let them be left,*
> *O let them be left, wildness and wet;*
> *Long live the weeds and the wilderness yet.*

The modern intellectual feels very superior when he contemplates the cosmology unquestioningly accepted by his ancestors. Their view of the universe was always so quaintly homocentric. Everything must find its ultimate explanation in man's needs. There was only God's will on the one hand and the natural world that he created for man's convenience on the other. But we have only exchanged one kind of homocentric cosmology for another. For God's will we have, to be sure, substituted the mindless mechanism of Darwinian evolution and assumed that a mere accident put us in a position to take advantage of what chance created. But in at least one respect we have only arrived by a different route at the same conclusion as our benighted ancestors: Only we count. Only we have minds. Only we can escape from absurdity. Only we, lifting ourselves by our own bootstraps, can save ourselves.

Pure Darwinism insists that we were elevated to what we regard as our exalted state by sheer accident followed by ineluctable necessity and that if nature seems to have been wise that is sheer illusion. But perhaps there was some wisdom as well as blind necessity in her processes. Man needs a context for his life larger than himself; he needs it so desperately that all modern despairs go back to the fact that he has rejected the only context which the loss of his traditional gods has left accessible. If there is any "somehow good" it must reside in nature herself. Yet the first item of our creed is a rejection of just that possibility.

"Nature books" generally get kindly reviews even from those who regard them as no more than the eccentric outpourings of harmless hobbyists. For that reason it was a salutary shock to come across in the admirable but not sufficiently well known magazine *Landscape* an excoriation of the whole tribe of nature writers in a review by Professor Joseph Slate of a book to which I had contributed a chapter. Its author calls us for the most part mere triflers in the genteel tradition, too devoted to the gentler pages of Thoreau and Muir, too little aware of the Dark Gods.

He has, to put it mildly, a point. We do not dig deep enough. We slip too easily into a spinsterish concern with the pretty instead of the beautiful. We tend to get only just far enough away from the "cute" to think of the natural world as a primarily gentle consolation instead of a great force, and we are content to experience it only superficially. He is also quite right when he accuses us of quoting too seldom those passages of Thoreau which face, as we hesitate to face, things dark enough in one way if not exactly in the way that D. H. Lawrence celebrated.

Consider, for example, the passage from Walden that begins: "We can never have enough of nature." Stop there and you might conclude that he was about to embark upon just the kind of discourse Professor Slate finds so objectionable. But read on and you will come in the next sentence to wildness at its most wild:

> *We can never have enough of nature. We must be refreshed by the sight of inexhaustible vigor, vast and titanic features, the seacoast with its wrecks, the wilderness with its living and its decayed trees, the thunder cloud, and the rain which lasts three weeks and produces freshets. We need to witness our own limits transgressed, and some life pasturing freely where we never wander. . . . I love to see that nature is so rife with life that myriads can be afforded to be sacrificed and suffered to prey on one another; that tender organizations can be so serenely squashed out of existence like pulp,—tadpoles which herons gobble up, and tortoises and toads run over in the road; and that sometimes it has rained flesh and blood! . . . Poison is not poisonous after all, nor are any wounds fatal.*

This is not the world we made and it is not the world we hope for. Neither is it the one where Thoreau himself most often felt at home. But it is at least an all but inexhaustible potentiality. Powerful as our weapons are, vast as is the destruction we are capable of, there is something still more powerful than we. That something is in part the least amiable but also the oldest and most enduring aspect of what Thoreau called "wildness," and it may survive when we have destroyed the better order we tried to make. By managing nature we may to some extent discipline it. We may also, in the process of becoming human, shift somewhat the emphases in its complex of impulses and powers. But we cannot dispense with the wildness without becoming near-machines and therefore less rather than more than the animal we try to transcend. Call it mother nature if you prefer a softer, less adequate term, but Thoreau preferred to call it wildness because he realized that the spectacle of wild nature reminds us so vividly of the fact that, though civilization is destructible and perhaps dispensable, the thing we civilize is not. To remind ourselves of its existence is only one (but not the least important) of the rewards for those who "in the love of Nature hold communion with her visible forms."

Thoreau was not a pessimist. He had faith in man's potentialities, though little respect for the small extent to which man had realized them. He did not believe, as some do today, that man was a failure beyond redemption, he believed only that his contemporaries were failures because the true way had been lost a long time ago and primitive man (still a savage even if a noble one) had taken the wrong road toward what he mistakenly believed to be wisdom and happiness. Thoreau's return to nature was a return to the fatal fork, to a road not taken, along which he hoped that he and others after him might proceed to a better future.

Please note that he called "wildness" merely the "preservation" of the world, not its "redemption" or "salvation." It may well be that its redemption may depend upon aspects of nature almost exclusively human. But man of the present day is more and more inclined to feel that mere survival or preservation is all he can hope for in the immediate future. If he is indeed granted a second chance to discover a genuinely good life, it may require him to go far back to that point where the road not taken branched off from the dubious road we have been following for so long and which we more and more stubbornly insist is the only right one because it takes us further and further away from the nature out of which we arose.

How far back would we have to go to find that road not taken? Could we, even supposing that we wanted to find it, reverse the direction in which our civilization is moving; or are the Marxians right and man is not a free agent but inevitably the captive of evolving technology that carries him along with it willy-nilly.

Thoreau thought he had an answer: "Simplify!" We get what we want and if things are in the saddle it is because we have put them there. And I have no doubt that he would say, "If you have hydrogen bombs that you are beginning to suspect you don't really want it is because you have for too long believed that power was the greatest of goods and did not realize soon enough that it would become, as it now has, a nemesis."

Perhaps he was right and perhaps if we don't have a change of heart, if we don't voluntarily simplify, we and our civilization will be simplified for us in a grand catastrophe. That would take us a good deal further back than is necessary to find the road not taken.

In the previous essay, Joseph Wood Krutch examined the general philosophical implications of Thoreau's statement that "In Wildness is the preservation of the World." Peter Matthiessen deals with a specific loss of wildness; man's exploitation of North American wildlife as evidenced by his extermination of the great auk over a century ago. Matthiessen is concerned with the meaning behind the needless extinction of this and other species. As William Beebe has said, ". . . when the last individual of a race of living things breathes no more, another heaven and another earth must pass before such a one can be again."

The Christian theology of Western man teaches him that all the creatures of the earth exist only for his benefit. Comfortable in this assurance, man in North America has left an awesome record of plunder upon his wildlife. The book Wildlife in America, *from which this is the opening chapter, is Matthiessen's examination of this record. Peter Matthiessen (1903—) has several other books, both nonfiction and fiction, to his credit.*

THE OUTLYING ROCKS

by Peter Matthiessen

These Penguins are as bigge as Geese, and flie not . . . and they multiply so infinitely upon a certain flat Iland, that men drive them from thence upon a boord into their Boates by hundreds at a time; as if God had made the innocencie of so poore a creature to become an admirable instrument for the sustenation of man.

—*RICHARD WHITBOURN (1618)*[1]

In early June of 1844, a longboat crewed by fourteen men hove to off the skerry called Eldey, a stark, volcanic mass rising out of the gray wastes of the North Atlantic some ten miles west of Cape Reykjanes, Iceland. On the islets of these uneasy seas, the forebears of the boatmen had always hunted the swarming sea birds as a food, but on this day they were seeking, for collectors, the eggs and skins of the garefowl or great auk, a penguin-like flightless bird once common on the ocean rocks of northern Europe, Iceland, Greenland, and the maritime provinces of Canada. The great auk, slaughtered indiscriminately across the centuries for its flesh, feathers, and oil, was vanishing, and the last birds, appearing now and then on lonely shores, were granted no protection. On the contrary, they were pursued more intensively than ever for their value as scientific specimens.

At the north end of Eldey, a wide ledge descends to the water, and, though a sea was running, the boat managed to land three men, Jon Brandsson, Sigourour Isleffson, and Ketil Ketilsson. Two auks, blinking, waddled foolishly across the ledge. Isleffson and Brandsson each killed a bird, and Ketilsson, discovering a solitary egg, found a crack in it and smashed it. Later, one Christian Hansen paid nine pounds for the skins, and sold them in turn to a Reykjavik taxidermist named Moller. It is not known what became of them thereafter, a fact all the more saddening when one considers that, on all the long coasts of the northern ocean, no auk was ever seen alive again.

The great auk is one of the few creatures whose final hours can be documented with such certainty. Ordinarily, the last members of a species die in solitude, the time and place of their passage from the earth unknown. One year they are present, striving instinctively to maintain an existence many thousands of years old. The next year they are gone. Perhaps stray auks persisted a few years longer, to die at last through accident or age, but we must assume that the ultimate pair fell victim to this heedless act of man.

One imagines with misgiving the last scene on desolate Eldey. Offshore, the longboat wallows in a surge of seas, then slides forward in the lull, its stem grinding hard on the rock ledge. The hunters hurl the two dead birds aboard and, cursing, tumble after, as the boat falls away into the wash. Gaining the open water, it moves off to the eastward, the rough voices and the hollow thump of oars against wood tholepins unreal in the prevailing fogs of June. The dank mist, rank with marine smells, cloaks the dark mass, white-topped with guano, and the fierce-eyed gannets, which had not left the crest, settle once more on their crude nests, hissing peevishly and jabbing sharp blue bills at their near neighbors. The few gulls, mewing aimlessly, circle in, alighting. One banks, checks its flight, bends swiftly down upon the ledge, where the last, pathetic generation of great auks gleams raw and unborn on the rock. A second follows and, squalling, they yank at the loose embryo, scattering the black, brown, and green shell segments. After a time they return to the crest, and the ledge is still. The shell remnants lie at the edge of tideline, and the last sea of the flood, perhaps, or a rain days later, washes the last piece into the water. Slowly it drifts down across the sea-curled weeds, the anchored life of the marine world. A rock minnow, drawn to the strange scent, snaps at a minute shred of auk albumen; the shell fragment spins upward, descends once more. Farther down, it settles briefly near a *littorina*, and surrounding molluscs stir dully toward the stimulus. The periwinkle scours it, spits the calcified bits away. The current takes the particles, so small as to be all but invisible, and they are borne outward, drifting down at last to the deeps of the sea out of which, across slow eons of the Cenozoic era, the species first evolved.

For most of us, its passing is unimportant. The auk, from a practical point of view, was doubtless a dim-witted inhabitant of Godforsaken places, a primitive and freakish thing, ill-favored and ungainly. From a second and a more enlightened viewpoint, the great auk was the mightiest of its family, a highly evolved fisherman and swimmer, an ornament to the monotony of northern seas, and for centuries a crucial food source for the natives of the Atlantic coasts. More important, it was a living creature which died needlessly, the first species native to North America to become extinct by the hand of man. It was to be followed into oblivion by other creatures, many of them of an aesthetic and economic significance apparent to us all. Even today, despite protection, the scattered individuals of species too long persecuted are hovering at the abyss of extinction, and will vanish in our lifetimes.

The slaughter, for want of fodder, has subsided in this century, but the fishes, amphibians, reptiles, birds, and mammals—the vertebrate animals as a group—are obscured by man's dark shadow. Such protection as is extended them too rarely includes the natural habitats they require, and their remnants skulk in a lean and shrinking wilderness. The true wilderness —the great woods and clear rivers, the wild swamps and grassy plains which once were the wonder of the world—has been largely despoiled, and today's voyager, approaching our shores through the oiled waters of the coast, is greeted by smoke and the glint of industry on our fouled seaboard, and an inland prospect of second growth, scarred landscapes, and sterile, often stinking, rivers of pollution and raw mud, the whole bedecked with billboards, neon lights, and other decorative evidence of mankind's triumph over chaos. In many regions the greenwood not converted to black stumps no longer breathes with sound and movement, but is become a cathedral of still trees; the plains are plowed under and the prairies ravaged by overgrazing and the winds of drought. Where great, wild creatures ranged, the vermin prosper.

The concept of conservation is a far truer sign of civilization than that spoliation of a continent which we once confused with progress. Today, very late, we are coming to accept the fact that the harvest of renewable resources must be controlled. Forests, soil, water, and wildlife are mutually interdependent, and the ruin of one element will mean, in the end, the ruin of them all. Not surprisingly, land management which benefits mankind will benefit the lesser beasts as well. Creatures like quail and the white-tailed deer, adjusting to man, have already shown recovery. For others, like the whooping crane, it is probably much too late, and the grizzly bear and golden eagle die slowly with the wilderness.

"Everybody knows," one naturalist has written, "that the autumn landscape in the north woods is the land, plus a red maple, plus a ruffed grouse. In terms of conventional physics, the grouse represents only a millionth of either the mass or the energy of an acre. Yet subtract the grouse and the whole thing is dead."[2]

The finality of extinction is awesome, and not unrelated to the finality of eternity. Man, striving to imagine what might lie beyond the long light years of stars, beyond the universe, beyond the void, feels lost in space; confronted with the death of species, enacted on earth so many times before he came, and certain to continue when his own breed is gone, he is forced to face another void, and feels alone in time. Species appear and, left behind by a changing earth, they disappear forever, and there is a certain solace in the inexorable. But until man, the highest predator, evolved, the process of extinction was a slow one. No species but man, so far as is known, unaided by circumstance or climatic change, has ever extinguished another, and certainly no species has ever devoured itself, an accomplishment of which man appears quite capable. There is some comfort in the notion that, however *Homo sapiens* contrives his own destruction, a few creatures will survive in that ultimate wilderness he will leave behind, going on about their ancient business in the mindless confidence that their own much older and more tolerant species will prevail.

The *Terra Incognita*, as cartographers of the Renaissance referred to North America, had been known to less educated Eurasians for more than ten thousand years. Charred animal bones found here and there in the West, and submitted to the radiocarbon test, have been ascribed to human campfires laid at least twenty-five thousand years ago. Thus one might say that the effect of man on the fauna of North America commenced with the waning of the glaciers, when bands of wild Mongoloid peoples migrated eastward across a land bridge now submerged by the shoal seas of the Bering Strait. In this period—the time of transition between the Pleistocene and Recent epochs—the mastodons, mammoths, saber-toothed tigers, dire wolves, and other huge beasts which had flourished in the Ice

Age disappeared forever from the face of the earth, and the genera which compose our modern wildlife gained ascendancy.

Man was perhaps the last of the large mammals to find the way from Asia to North America. In any case, many species had preceded him. The members of the deer family—the deer, elk, moose, and caribou—had made the journey long before, as had the bison, or buffalo, and the mountain sheep. Among all modern North American hoofed mammals, in fact, only the pronghorn antelope emerged originally on this continent. The gray wolf, lynx, beaver, and many other animals also have close relations in the Old World, so close that even today a number of them—the wolverine and the Eurasian glutton, for example, and the grizzly and Siberian brown bear—are widely considered to be identical species. Similarly, many bird species are common to both continents, including the herring gull, golden plover, mallard, and peregrine falcon. The larger groupings—the genera and families which contain those species and many others—are widespread throughout the Northern Hemisphere. Even among the songbirds, which are quite dissimilar on the two continents in terms of individual species, the only large American family which has no counterpart in Eurasia is that of the colorful wood warblers, *Parulidae.*

Since the American continents are connected overland, it seems rather strange that the faunas of North America and Eurasia are more closely allied than the faunas of North and South America. One must remember, however, that the Americas were separated for fifty million years or more in the course of the present geologic era, and during this time their creatures had evolved quite differently. It is only in recent times, in geological terms—two million years ago, perhaps—that the formation of the huge icecaps, lowering the oceans of the world, permitted the reappearance of the Panama bridge between Americas.

The animals moved north and south across this land bridge, just as they had moved east and west across the dry strait in the Arctic. But the South American forms, become senile and over-specialized in their long period of isolation, were unable to compete with the younger species which were flourishing throughout the Northern Hemisphere. Many archaic monkeys, marsupials, and other forms were rapidly exterminated by the invaders. Though a certain interchange took place across the land bridge, the northern mammalian genera came to dominate both continents, and their descendants comprise virtually all the large South American animals of today, including the cougar, jaguar, deer, peccaries, and guanacos.

The armadillo, opossum, and porcupine, on the other hand, are among the primitive creatures which arrived safely from the opposite direction and are still extending their range. A large relation of the armadillo, *Boreostracon*, and a mighty ground sloth, *Megatherium*, also made their way to North America. These slow-witted beasts penetrated the continent as far as Pennsylvania, only to succumb to the changes in climate which accompanied the passing of the Ice Age.

The mass extermination of great mammals at this time occurred everywhere except in Africa and southern Asia. Alteration of environment brought about by climatic change is usually held accountable, but the precise reasons are as mysterious as those offered for the mass extinction of the dinosaurs some seventy million years before. Even among large animals the extinctions were by no means uniform: in North America the moose and bison were able to make the necessary adaptations, while the camel and horse were not. The camel family survived in South America in the wild guanaco and vicuña, but the horse was absent from the Western Hemisphere until recent centuries, when it returned with the Spaniards as a domestic animal.

Large creatures of the other classes were apparently less affected than the mammals. Great Pleistocene birds such as the whooping crane and the California condor prevail in remnant populations to this day, and many more primitive vertebrates, of which the sharks, sturgeons, sea turtles, and crocodilians are only the most spectacular examples, have persisted in their present form over many millions of years. For these, the slow wax and wane of the glacial epoch, which witnessed the emergence of mankind, was no more than a short season in the long history of their existence on the earth.

The last mastodons and mammoths were presumably hunted by man, who may have been hunted in his turn by *Smilodon*, the unsmiling saber-toothed tiger. It is very doubtful, however, whether the demise of these creatures at the dawn of the Recent epoch was significantly hastened by nomadic hunters of the Eskimo, Athabascan, Iroquoian, Siouan, and Algonquian races, the numerous tribes of which were wandering east and south across the continent. The red men were always few in number and, the Pueblo peoples of the Southwest excepted, left little sign of their existence. They moved softly through the wilderness like woodland birds, rarely remaining long enough in one locality to mar it.

The visits by Vikings, few records of which have come down from the Dark Ages, were transient also, and the forest green soon covered their crude settlements, leaving only a few much-disputed traces. These fierce warriors, whose sea-dragon galleys were the most exotic craft ever to pierce the North Atlantic fogs, had colonized Greenland by the tenth century and were thus the earliest white discoverers of the Western Hemisphere. That they also discovered North America by the year 1000 seems hardly to be doubted, and the Norse colonists of an ill-defined stretch of northeast coast were the first to record the resources of the new continent. In addition to the wild grapes for which the country was called Vinland, "there was no lack of salmon there either in the river or in the lake, and larger salmon than they had ever seen before," according to the chronicle of Eric the Red.[3] But they concerned themselves chiefly with the export of timber and fur, and in their murderous dealings with the Skrellings, as they called the red men, established a precedent firmly adhered to in later centuries by more pious invaders from France, England, Spain, and Holland. The last Vinland colony, in 1011, was beset less by Skrellings than by civil strife; in the following spring, the survivors sailed away to Greenland, and the history of Vinland, brief and bloody, came to an end.

The modern exploitation of North American wildlife, then, commenced with Breton fishermen who, piloting shallops smaller still than the very small *Santa Maria*, were probably appearing annually on the Grand Banks off Newfoundland before the voyage of Columbus, and certainly no later than 1497, the year that Americus Vespucius and the Cabots explored Vinland's dark, quiet coasts. "The soil is barren in some places," Sebastian Cabot wrote of Labrador or Newfoundland, "and yields little fruit, but it is full of white bears, and stags far greater than ours. It yields plenty of fish, and those very great, as seals, and those which commonly we call salmons: there are soles also above a yard in length: but especially there is great abundance of that kind of fish which the savages call baccalaos."[4] The baccalao, or cod, abounding in the cold offshore waters of the continental shelf, formed the first major commerce of what Vespucius, in a letter to Lorenzo de' Medici, would term the New World; in its incidental persecution of sea birds, this primitive fishery was to initiate the long decline of North American fauna.

Though the Breton fishermen left no records, it must be assumed that they located almost immediately the great bird colonies in the Magdalen Islands and at Funk Island, a flat rock islet thirty-odd miles off Newfoundland. Since many sea birds, and especially those of the alcid family—the auks, puffins, guillemots, and murres—are of general distribution on both sides of the North Atlantic and nest on the rock islands of Brittany even today, these sailors were quick to recognize their countrymen. A concept of the plenty they came upon may still be had at Bonaventure Island, off the Gaspé Peninsula of Quebec, where the four-hundred-foot cliffs of the seaward face form one vast hive of alcids. The birds swarm ceaselessly in spring and summer, drifting in from the ocean in flocks like long wisps of smoke and whirring upward from the water to careen clumsily along the ledges. Above, on the crest, the magnificent white gannets nest, and the kittiwakes and larger gulls patrol the face, their sad cries added to a chittering and shrieking which pierce the booming of the surf in the black sea caves below. At the base of the cliff the visitor, small in a primeval emptiness of ocean, rock, and sky, feels simultaneously exalted and diminished; the bleak bird rocks of the northern oceans will perhaps be the final outposts of the natural profusion known to early voyagers, and we moderns, used to remnant populations of creatures taught to know their place, find this wild din, this wilderness of life, bewildering.

The largest alcid, and the one easiest to kill, was the great auk. Flightless, it was forced to nest on low, accessible ledges, and with the white man's coming its colonies were soon exterminated except on remote rocks far out at sea. The size of a goose, it furnished not only edible eggs but meat, down and feathers, oil, and even codfish bait, and the Micmac Indians were said to have valued its gullet as a quiver for their arrows. The greatest colony of garefowl was probably at Funk Island, where Jacques Cartier, as early as 1534, salted down five or six barrels of these hapless birds for each ship in his expedition. In 1536 an Englishman named Robert Hore improved upon old-fashioned ways by spreading a sail bridge from ship to shore and marching a complement of auks into his hold. Later voyagers, sailing in increasing numbers to the new continent, learned quickly to augment their wretched stores in similar fashion, not only at Funk Island but at Bird Rocks in the Magdalens and elsewhere. The great auk is thought to have nested as far south as the coast of Maine, with a wintering population in Massachusetts Bay, but the southern colonies were probably destroyed quite early.

As a group, the alcids have always been extraordinarily plentiful—the Brünnich's murre and the dovekie, which may be the most numerous of northern sea birds, each boast colonies in Greenland of two million individuals or more—and the great auk was no exception. The relative inaccessibility of its North Atlantic rookeries deferred its extinction for three centuries, but by 1785, when the frenzy of colonization had subsided, George Cartwright of Labrador, describing the Funks, was obliged to take note of the bird's decline: ". . . it has been customary of late years, for several crews of men to live all summer on that island, for the sole purpose of killing birds for the sake of their feathers, the destruction which they have made is incredible. If a stop is not soon put to that practice, the whole breed will be diminished to almost nothing, particularly the penguins: for this is now the only island they have left to breed upon. . . ."[5] Cartwright does not mention the complementary industry of boiling the birds in huge try-pots for their oil, an enterprise made feasible on the treeless Funks by the use of still more auks as fuel.

The naturalists of the period, unhappily, did not share Cartwright's alarm. Thomas Pennant, writing in the previous year, makes no mention of auk scarcity, and Thomas Nuttall, as late as 1834, is more concerned with the bird's demeanor than with its destruction. "Deprived of the use of wings," he mourns, "degraded as it were from the feathered ranks, and almost numbered with the amphibious monsters of the deep, the Auk seems condemned to dwell alone in those desolate and forsaken regions of the earth. . . . In the Ferröe isles, Iceland, Greenland and Newfoundland, they dwell and breed in great numbers. . . . [6] Though Nuttall pointed out, somewhat paradoxically, that recent navigators had failed to observe them, his contemporary, Mr. Audubon, was persuaded of their abundance off Newfoundland and of their continued use as a source of fish bait. In 1840, the year after Audubon's account, the auk is thought to have become extinct off Newfoundland, and two decades later Dr. Spencer F. Baird was of the opinion that, as a species, the bird was rather rare. His remark may well have been the first of a long series of troubled observations by American naturalists in regard to the scarcity of a creature which was, in fact, already extinct.

"All night," wrote Columbus, in his journal for October 9, 1492, "they heard birds passing."[7] He was already wandering the eastern reaches of the Caribbean, seeking in every sign of life a harbinger of land. The night flyers mentioned were probably hosts of migratory birds, traversing the Caribbean from North to South America, rather than native species of the Greater Antilles or Hispaniola. Columbus could not have known this, of course, nor did he suspect that the birds seen by day which raised false hopes throughout the crossing were not even coastal species, but shearwaters and petrels, which visit land but once a year to breed.

Certain shearwaters, storm petrels, and alcids are still very common in season off the Atlantic coasts, but it is no coincidence that the great auk and two species of petrel were the first North American creatures to suffer a drastic decline. The Atlantic islands, rising out of the endless fetch of the wide, westward horizon, were much frequented by ships, and often provided new ship's stores for the last leg of the voyage. Fresh meat was usually supplied by sea birds, incredibly plentiful on their crowded island nesting grounds; in temperate seas, the shearwaters and petrels, like the great auks farther north, were conscripted commonly as a supplementary diet.

In spite of local plenty, the bird communities of islands around the world are often early victims of extermination. The breeding range of island species is small and therefore vulnerable, and the species themselves may be quite primitive. Some are relict populations of forms which, on the mainland, have long since succumbed in the struggle for survival. Other species, freed from competition and mammalian predation, grow overspecialized, diminished in vitality, and thus are ill equipped to deal with new factors in their environment.

Man is invariably a new factor of the most dangerous potential. The fiercest animal of all, he is especially destructive when he introduces, in addition to himself, such rapacious mammal relatives as the rat, the mongoose, and the cat, all of them beasts superbly equipped to make short work of birds, eggs, and other edible life escaping the attention of their large ally.

The ship rat may have explored Bermuda as early as 1603. That year a Spanish crew under Diego Ramírez, frightened at first by the unearthly gabblings of myriad nocturnal spirits, discovered upon closer inspection that these evil things were birds, and highly palatable birds at that. The good impression of the Spaniards was confirmed six years later by a Mr. W. Strachey, shipwrecked in those parts with Sir George Somers on the *Sea Venture*,[8] who wrote as follows:

> *A kind of webbe-footed Fowle there is, of the bigness of an English greene Plover, or Sea-Meawe, which all the Summer we saw not, and in the darkest nights of November and December . . . they would come forth, but not flye farre from home, and hovering in the ayre, and over the Sea, made a strange hollow and harsh howling. They call it of the cry which it maketh, a cohow. . . . There are thousands of these Birds, and two or three Islands full of their Burrows, whether at any time . . . we could send our Cockboat and bring home as many as would serve the whole Company.*[9]

Strachey's implication that the nesting burrows were confined to a few islands—or more properly, islets—is significant, for the cahow, or Bermuda petrel, was the first New World example of a creature endangered by its narrow habitat. The cahow's original nesting range throughout the islands was doubtless restricted to the offshore rocks not long after the first sail broke the ocean horizons. In addition to man and his faithful rats, a number of hogs were turned loose in the Bermudas very early, and these are thought to have rooted out the colonies on the larger islands. Nevertheless, the cahow and the "pimlico," known today as Audubon's shearwater, remained abundant on the islets of Castle Roads and elsewhere, and it may have been the famine in the winter of 1614-1615 which brought about the final decline of the former. The following year, a proclamation was issued "against the spoyle and havock of the Cahowes, and other birds, which already wer almost all of them killed and scared away very improvidently by fire, diggeing, stoneing, and all kinds of murtherings." A law protecting the nesting birds was passed in 1621 which, to judge from its results, was unavailing. About 1629, scarcely a quarter-century after the first accounts of it, the cahow disappeared entirely.

In the ordinary course of events, the cahow would have thus become the first North American species to die by the hand of man. (Bermuda is here considered an extension of North America, since it cannot be geographically allied to any other land mass and since, in this period, it was part of the Virginia Colony. For the purposes of this [article], North America may be taken to include the continent north of the Mexican border, with its offshore islands and the oceanic islands of Bermuda, although the border is not a continental line, and is somewhat north of the vague faunal "boundary" which roughly separates the representative animals of the two Americas.) But the species was marvelously resurrected in 1906, when an unknown petrel was discovered in a Castle Island crevice. The bird at first was considered a new species, but three other specimens located in subsequent years closely fitted a description of the historic cahow constructed from antique remains. In 1951, some nesting burrows, occupied, were found on islets near Castle Roads. Carefully guarded, these burrows are nonetheless subject to the whims of rats as well as to confiscation by the yellow-billed tropic birds, and there is small hope that the cahow's stamina can maintain it another century. Less than one hundred individuals are now thought to exist, but the fact remains that the species managed to survive nearly three hundred years of supposed extinction. Its status as a "living fossil" cannot compare with that of the coelacanth which, first captured off

South Africa in 1938, is a five-foot specimen of an order of fossil fishes thought to have vanished from the earth, not three hundred but three hundred million years ago. Nevertheless, the cahow's story is remarkable, and one must admire the persistence of the survivors. Scattered out across the great Atlantic, they have homed to their rock islets every autumn, year after year after year, to perpetuate their kind beneath the very shadows of the planes which fly man in and out of Bermuda's airfield.

Though remains have been found in the Bahamas, the former occurrence of the cahow off the coasts of the Atlantic States can only be presumed. Similarly, the Guadalupe petrel, first noted on Guadalupe Island off Baja California in 1887, has never been recorded elsewhere, though probably it ranged to California before cats left behind by transient fishermen apparently overpowered it. Little is known of its original distribution, and it is not likely that we will learn much more, the species having disappeared after 1912. The short-tailed albatross of the western Pacific, on the other hand, was sighted offshore commonly, from Alaska to California, until an Oriental market for its feathers all but finished it, and is therefore a member *in absentia* of our fauna.

The diablotin, or black-capped petrel, not only has visited our coasts but has journeyed far inland. This oceanic wanderer makes its nest in West Indian mountain burrows, and has turned up, usually after storms, in such unlikely haunts as Kentucky, Ohio, New Hampshire, Ontario, and Central Park, in New York City. As a significant food source for mankind, however, it has a history almost as dark and brief as that of its near-relative, the cahow. An account dating from 1696 refers probably to the diablotin in observing that "The difficulty of hunting these birds preserves the species, which would have been entirely exterminated years ago, according to the bad custom of the French, did they not retire to localities which are not accessible to everyone."[10] Localities inaccessible to man, however, were readily accessible to mongooses and opossums imported to its islands, and as a consequence the species has all but disappeared. Its last nesting grounds are unknown, and the few sightings of this large black-and-white petrel in recent decades are largely of random individuals glimpsed on Columbus's western ocean. Columbus himself may well have seen it, and the ultimate record will doubtless be made from aboard ship. One imagines with a sense of foreboding this strange, solitary bird passing astern, its dark, sharp wing rising and vanishing like a fin as it banks stiffly among the crests until, scarcely discernible, it fades into eternities of sea.

The decision to build a dam on a free-flowing river is usually endorsed by local politicians and businessmen, their eyes on the construction payroll, and by the Army Corps of Engineers, whose principle reason for existence has been to build dams. Sometimes, of course, support is more widespread if the dam promises to solve serious problems of water shortage and flooding, and in such cases there can be no argument with the decision to build a dam. But often, as this essay demonstrates, dams are built where they are not really needed and where their construction means permanent and immeasurable environmental loss. Harry Caudill (1922-) laments the planned destruction of a Kentucky wilderness region of incomparable beauty. Because of such examples of heedless damming and because so many of our wild rivers have already been reduced to a series of standing pools, William O. Douglas (page 301) and others have called for the Corps of Engineers to turn their attention from building dams to designing and constructing waste disposal plants and the like.

Harry Caudill is a Kentucky attorney and former state legislator. His concern for the natural environment of his state has led him to oppose not only the damming of the Red River but also the destruction of the southern Appalachians by coal mining interests.

A WILD RIVER THAT KNEW BOONE AWAITS ITS FATE

by Harry M. Caudill

The gorge of the Red River of Kentucky—in many, perhaps too many ways the little Grand Canyon of eastern America—is about to die. It will drown beneath the rising waters of yet another flood control dam. And in its extinction all who love beauty and nature today, and all who will cherish them tomorrow, will sustain immeasurable loss. For the Red River Valley is not merely pretty. It is magnificent. It is lovely to a degree difficult to describe. The beauty of the place varies with its gentle moods. The charm of the ancient chasm is altered with each hour, indeed, with each cloud that casts a shadow on its gray crags. No greater condemnation could be written of this generation than that it callously condemned this valley and all its ancient and varied forms of life. Tiny compared with the Grand Canyon of the Colorado, it has a character as haunting, mysteries as intriguing, enchantments as subtle. Unknown to the nation until recently, it has been sentenced to death at the very moment of its discovery.

The Red River is as thoroughly Kentuckian as the very name of the commonwealth. Rising near the Magoffin County line, it murmurs westward 50 miles as the crow flies and twice as far if all its convolutions are measured.

"A Wild River That Knew Boone Awaits Its Fate" Reprinted from *Audubon*, September-October, 1968.

For a quarter of its length it tumbles between sandstone escarpments and past eroded, tree-grown mesas and chimneys. Crystalline for most of the year, the water laps against boulders fallen from the beetling cliffs and, for long stretches, lies in limpid pools before draining away over rocky shallows. Where Swift Camp Creek and other tributaries tumble into the river, leaves eddy along the edges of deep holes, while lichened cliffs tell of an incomprehensible antiquity. And well they might, for the Red River has chiseled at these sandstone crests and the white limestone at their foundations for 60 millions of years.

But water alone has not carved a winding trench nearly 600 feet into the rock. In the spring and on still summer days, the wind whispers along the cliffs. But in wintry storms it howls and it claws at the stone, aided by the leaden weight of ice that sweeps downward like the beards of titanic patriarchs. The wind and ice and changing temperatures tear grains and slabs from the retreating walls, carving caves, spires, domes, arches. More than thirty natural bridges span rivers of wind and sky. One is a hundred feet long and soars twenty-five feet to a graceful apex.

Most of the canyon lies within the 460,577-acre Daniel Boone National Forest, where the timber has escaped the ax for a third of a century. In scores of deep, shadowed hollows the tulip poplars, hemlocks, oaks, hickories, and beeches rise straight and unblemished, telling, in some measure at least, of the splendor that must have marked these forests when Boone and other long hunters first penetrated what they called the "roughs of the Red River."

That Daniel Boone often hunted deer and black bear in the roughs cannot be doubted. A day's walk from the edge of the Bluegrass, the gorge was a favored hunting ground for generations after game had disappeared from the "plains." A few years ago foresters entered a high, overhanging cliff or rockhouse where the sandy floor had lain dry and undisturbed for many decades. There they discovered a crude hut built of split white oak boards and set like a small fortress a yard deep in the earth. Nearby was a trough fashioned from a hollowed poplar log and two small furnaces of piled stones that showed evidence of having endured great heat—perhaps in the smelting of lead for bullets. Within the shelter was a board burned with the letters D. BooN. Historians from the University of Kentucky agree that the lettering is very old, that it resembles the great scout's, and that it is probably a memento he left behind when he broke camp one misty morning nearly two centuries ago.

The fire-softened furnaces at the Boone camp set off a flurry of new interest in the venerable tradition of the lost silver mine. Swift Camp Creek does not take its name from the fast-flowing water in its channel but from John Swift, who braved the threat of Shawnee torture to prospect for precious metals and, according to legend, found a rich lode of silver on a creek in eastern Kentucky. Local versions put the strike in the gorge along a tributary creek flowing in from the south. But Swift fell on evil days. His companions murdered one another or were slain by Indians. Swift stumbled out of the wilderness sick and blind, and was never able to lead a search party to his bonanza. Today one may still see an "Indian stairway"—toe and finger holds carved across a cliff. Although archaeologists believe they were carved by Indians, Wolfe County natives cling to the account handed down from their forebears that they were cut by Swift in his search for outcroppings of the coveted metal.

In any event, Boone and his long hunters were late visitors. Last summer, archaeological teams digging under cliffs and in the narrow bottomlands found potsherds, flints, and other artifacts at many layers. They also unearthed traces of ancient palisaded villages, indicative of long-established community life. The scientists believe the Red River Gorge was inhabited 4,000 years ago, that for centuries it supported and sheltered a considerable copper-colored population. Nor has the game that attracted Boone and other long hunters wholly disappeared from its streams, coves, and thickets.

The water holes of the Red River remain deep and clear except during sustained drought, and such intervals are rare in the Kentucky mountains. In the gorge the river becomes white water, potable for many miles. (Unfortunately, cleared land on the headwaters has begun to release mud that could eventually impair the stream.) Shaded by willows, birches, beeches, and sycamores, and fringed for long distances with wild cane, the Red River and its larger tributaries are the cool and pleasant habitat of many of North America's finest gamefish. In spring, bass, perch, muskie, trout, and catfish lie on beds of smooth jackstones or linger in the shadows of boulders. In the autumn bass break the surface to catch insects swarming in the waning sunlight. It is hard for a trout fisherman to ply the Red River without creeling his limit, while escaping the burning sun and the raucous outboards which make lake fishing less pleasant recreation.

In a two-mile hike up Swift Camp Creek, my wife, Anne, and I discovered three beaver dams, each with an array of driftwood behind it. We met a trapper returning heavy-laden from a fortnight's solitary labor along his trapline. Only the night before he had caught a 40-pound beaver, and he showed us an impressive collection of pelts taken earlier. There were, in addition to beaver, gray and red fox, muskrat, raccoon, and weasel. He assured us there were deer, black bear, and bobcats living in the valley, and he ventured the opinion that, of the canyon's original fauna, only mountain lions and wolves had disappeared. He had breakfasted on trout caught on a baited setline while he slept.

The mast fall from walnut, oak, hickory, and beech is generous, and the gorge and the many spurs and points that jut into it resound at dawn with the chatter and squeak of gray and red squirrels. But on our many trips to the gorge we have seen few songbirds, although we have frequently sighted hawks, crows, and vultures. At least one ornithologist reports having seen golden eagles nesting in the top of a dead pine. Biologists claim to have recorded 274 other species, resident and transient.

But it is the ground life, the profusion of startlingly colorful wild flowers and clumps of fern, that brings the gorge to glory in late April and early May. Within 1,000 feet, as one moves up from the floor of the canyon to the dry, pine-grown, sand-capped ridgetop, one meets many of the climatic conditions prevailing between the Canadian border and northern Georgia. From the fern-rimmed rivulets, up through the moist, poplar-studded coves, in crevices, on ledges of sheer cliffs, and amid layered pine needles on the dry mesa tops, all hues are encountered. The gentians, moccasin flower, partridge-berry, Indian pipe, trout lily, Turk's-cap lily, azalea, cardinal flower, trillium, groundsel, goldenrod, watercress, sage, Dutchman's breeches, mountain-mint—these and many others riot in red, purple, white, yellow, all shadings in between.

It is ironic, then, that the doom of this lovely corner of America has come in a time of mushrooming population, swelling demands for the preservation of wild rivers, a White House Conference on Natural Beauty, and tree plantings by the First Lady.

Lexington, Kentucky, is an hour's drive from the Red River Gorge. Once a lovely old town, it is being transformed into an ordinary, crowded, dirty, overbuilt industrial city. It claims to need reliable new sources of water (four years ago it had such a surplus that the Lexington Water Company sold two of its reservoirs). And the U.S. Army Corps of Engineers has convinced the city that the Red River is an ever-flowing fount, just waiting to be tapped.

Occasionally, too, the Red River has left its banks and flooded the little village of Clay City. Thus, it is said, a dam on the Red River will keep Clay City dry while keeping Lexington properly wet. The mayor and aldermen thus clamored for the project to get underway. No alternatives were considered. Bluegrass Congressman John Watts came to the project's support; he is a power to be reckoned with because he sits on the supremely important House Ways and Means Committee. The project received congressional sanction and an initial appropriation of $581,000, and President Johnson included $760,000 in the 1968-1969 budget to continue the project.

Local Jaycees and civic boosters in and around Lexington are, of course, overjoyed with the thought of a new lake. But not all are so pleased.

Chester and Dorothy Morrison live on a farm in the gorge, their cottage and fields surrounded by towering cliffs and stands of tall timber. No neighbor is within sight. Their children live in an Eden of fresh air, sparkling water, open spaces, clear skies. The Morrisons are desolate because they will have to leave the valley.

In December, Supreme Court Justice William O. Douglas led a protest hike through the canyon. Six hundred persons from several states accompanied him, and later deluged the White House and Congress with letters urging that the valley be spared.

Under the leadership of Professor James Kowalski of Union College at Barbourville, the Kentucky section of the Sierra Club fought a valiant battle for the canyon and, despite hard and numerous setbacks, continued to plead with Washington for a change of heart.

The Louisville Courier-Journal constantly defended the gorge in articles and editorials. It enlisted the aid of the Kentucky Federation of Garden Clubs, the Kentucky Federation of Women's Clubs, the Louisville and Buckley Hills Audubon Societies, and the Kentucky Academy of Science. Congressional mail ran as much as ten-to-one against the reservoir.

The prestigious American Association for the Advancement of Science also entered the lists on behalf of the gorge. Its council unanimously condemned the proposed dam on the ground that it would destroy "a unique natural area of geological, botanical, and zoological significance, valuable in research and instruction."

And certain officers in the Corps of Engineers even confided that the project deserved thorough reconsideration. At a public hearing called by former Governor Edward Breathitt, Corps spokesmen conceded that a dam at an alternate site farther downstream would save most of the gorge without substantially lessening the "benefits." Ignored was the opinion of competent engineers that any dam will be useless because the limestone at the valley floor is cracked and porous and will not hold water.

A reprieve, indeed, for awhile seemed possible. Even though it had begun buying land, had spent $100,000 on engineering and design, and had obligated more thousands of dollars to initiate construction, the Corps of Engineers agreed to a restudy after that protest "march."

Perhaps, it was hoped, the Corps would be overwhelmed by the enormity it was about to commit.

But the Red River Gorge is not likely to be spared.

Its ultimate fate is lost in the confusing language of a congressional conference report. First, its "restudy" completed as a sop to its public conscience, the Corps said it would proceed with the dam as planned. It promptly received its money, its pork, from the President, the Bureau of the Budget, and, despite Senate objections, from Congress.

The Engineers have been told to try to preserve the gorge "to the maximum extent feasible," perhaps with a reservoir a bit smaller. But an alternate, somewhat less destructive dam site downstream has been rejected, and construction will probably begin early next year.

Dams, it is time those who love beauty (and sanity) must learn, are far more important to America than flowers and ferns, trees and legends, gorges and wild-running rivers.

Editors' note: Because of such articles as this and because of the continued protests of the individuals and groups mentioned herein, the planned dam on the Red River has now been moved downstream five miles so that most of the gorge will be preserved in its natural state.

Some
Philosophies
for Survival

The essays which comprise this final section point to some ways in which the mistakes of the past, as revealed in the earlier sections, can perhaps be corrected or averted in the future. It is appropriate to begin here with a selection from Science and Survival *by Barry Commoner (1917-) since he confronts the elemental dilemma of modern science: Science apparently gives us the power to advance our own well-being while at the same time it also threatens our continued survival. In this, the final chapter of his book, Commoner asks how scientist and citizen can together avoid the threats which accompany scientific advancement. Like Garrett Hardin, he asks us to weigh the benefits of each new technological innovation against the possible harm it could cause. For the future, he claims, the ecological effects of scientific innovations must be determined in advance rather than after the damage has been done, and he recognizes that even now we are ignoring this lesson. For example, several countries race to build a supersonic transport whose hazards have not yet been evaluated, and, in the ultimate of absurdities, a nuclear arms race continues in the face of universal acknowledgment that it will wreak catastrophic effects upon the earth and its inhabitants.*

Barry Commoner, often called "The Paul Revere of Ecology," is one of a new breed of scientists who concern themselves with the moral and social problems which science has created.

TO SURVIVE ON THE EARTH

by Barry Commoner

Once the problems are perceived by science, and scientists help citizens to understand the possible solutions, what actions can be taken to avoid the calamities that seem to follow so closely on the heels of modern technological progress? I have tried to show that science offers no "objective" answer to this question. There is a price attached to every solution; any judgment will necessarily reflect the value we place on the benefits yielded by a given technological advance and the harm we associate with its hazards. The benefits and the hazards can be described by scientific means, but each of us must choose that balance between them which best accords with our own belief of what is good—for ourselves, for society, and for humanity as a whole.

In discussing what ought to be done about these problems, I can speak only for myself. As a scientist, I can arrive at my own judgments—subject to the open criticism which is so essential to scientific discourse—about the scientific and technological issues. As a citizen, I can decide which of the alternative solutions my government ought to pursue, and, using the instruments of politics, act for the adoption of this course. As a human being, I can express in this action my own moral convictions.

As a biologist, I have reached this conclusion: we have come to a turning point in the human habitation of the earth. The environment is a complex, subtly balanced system, and it is this integrated whole which receives the impact of all the separate insults inflicted by pollutants. Never before in the history of this planet has its thin life-supporting surface been subjected to such diverse, novel, and potent agents. I believe that the cumulative effects of these pollutants, their interactions and amplification, can be fatal to the complex fabric of the biosphere. And, because man is, after all, a dependent part of this system, I believe that continued pollution of the earth, if unchecked, will eventually destroy the fitness of this planet as a place for human life.

My judgment of the possible effects of the most extreme assault on the biosphere—nuclear war—has already been expressed. Nuclear war would, I believe, inevitably destroy the economic, social, and political structure of the combatant nations; it would reduce their populations, industry and agriculture to chaotic remnants, incapable of supporting an organized effort for recovery. I believe that world-wide radioactive contamination, epidemics, ecological disasters, and possibly climatic changes would so gravely affect the stability of the biosphere as to threaten human survival everywhere on the earth.

If we are to survive, we need to become aware of the damaging effects of technological innovations, determine their economic and social costs, balance these against the expected benefits, make the facts broadly available to the public, and take the action needed to achieve an acceptable balance of benefits and hazards. Obviously, all this should be done *before* we become massively committed to a new technology. One of our most urgent needs is to establish within the scientific community some means of estimating and reporting on the expected benefits and hazards of proposed environmental interventions *in advance*. Such advance consideration could have averted many of our present difficulties with detergents, insecticides, and radioactive contaminants. It could have warned us of the tragic futility of attempting to defend the nation's security by a means that can only lead to the nation's destruction.

We have not yet learned this lesson. Despite our earlier experience with nondegradable detergents, the degradable detergents which replaced them were massively marketed, by joint action of the industry in 1965, without any pilot study of their ecological effects. The phosphates which even the new detergents introduce into surface waters may force their eventual withdrawal. The United States, Great Britain, and France are already committed to costly programs for supersonic transport planes but have thus far failed to produce a comprehensive evaluation of the hazards from sonic boom, from cosmic radioactivity, and from the physiological effects of rapid transport from one time zone to another.[1] The security of every nation in the world remains tied to nuclear armaments, and we continue to evade an open public discussion of the basic question: do we wish to commit the security of nations to a military system which is likely to destroy them?

It is urgent that we face this issue openly, now, before by accident or design we are overtaken by nuclear catastrophe. U Thant has proposed that the United Nations prepare a report on the effects of nuclear war and disseminate it throughout the world. Such a report could become the cornerstone of world peace. For the world would then know that, so long as nuclear war remains possible, we are all counters in a colossal gamble with the survival of civilization.

The costs of correcting past mistakes and preventing the threatened ones are already staggering, for the technologies which have produced them are now deeply embedded in our economic, social, and political structure. From what is now known about the smog problem, I think it unlikely that gasoline-driven automobiles can long continue to serve as the chief vehicle of urban and suburban transportation without imposing a health hazard which most of us would be unwilling to accept. Some improvement will probably result from the use of new devices to reduce emission of waste gasoline. But in view of the increasing demand for urban transportation any really effective effort to reduce the emission of waste fuel, carbon monoxide, and lead will probably require electric-powered vehicles and the replacement of urban highway systems by rapid transit lines. Added to current demands for highway-safe cars, the demand for smog-free transportation is certain to have an important impact on the powerful and deeply entrenched automobile industry.[2]

The rapidly accelerating pollution of our surface waters with excessive phosphate and nitrate from sewage and detergents will, I believe, necessitate a drastic revision of urban waste systems. It may be possible to remove phosphates effectively by major modifications of sewage and water treatment plants, but there are no methods in sight that might counter the accumulation of nitrate. Hence, control will probably need to be based chiefly on preventing the entry of these pollutants into surface waters.

According to a report by the Committee on Pollution, National Academy of Sciences,[3] we need to plan for a complete transformation of urban waste-removal systems, in particular to end the present practice of using water to get rid of solid wastes. The technological problems involved are so complex that the report recommends, as an initial step, the construction of a small pilot city to try out the new approach.

The high productivity of American agriculture, and therefore its economic structure, is based on the use of large amounts of mineral fertilizer in which phosphate and nitrate are major components. This fertilizer is not entirely absorbed by the crops and the remainder runs off into streams and lakes. As a result, by nourishing our crops and raising agricultural production, we help to kill off our lakes and rivers. Since there is no foreseeable means of removing fertilizer runoff from surface waters, it will become necessary, it seems to me, to impose severe restrictions on the present unlimited use of mineral fertilizers in agriculture. Proposed restraints on the use of synthetic pesticides have already aroused a great deal of opposition from the chemical industry and from agriculture. Judged by this response, an attempt to regulate the use of mineral fertilizers will confront us with an explosive economic and political problem.[4]

And suppose that, as it may, the accumulation of carbon dioxide begins to threaten the entire globe with catastrophic floods. Control of this danger would require the modification, throughout the world, of domestic furnaces and industrial combustion plants—for example, by the addition of devices to absorb carbon dioxide from flue gases. Combustion-driven power plants could perhaps be replaced with nuclear ones, but this would pose the problem of safely disposing of massive amounts of radioactive wastes and create the hazard of reactor accidents near centers of population. Solar power, and other techniques for the production of electrical power which do not require either combustion or nuclear reactors, may be the best solution. But here too massive technological changes will be needed in all industrial nations.

The problems of industrial and agricultural pollution, while exceedingly large, complex, and costly, are nevertheless capable of correction by the proper technological means. We are still in a period of grace, and if we are willing to pay the price, as large as it is, there is yet time to restore and preserve the biological quality of the environment. But the most immediate threat to survival—nuclear war—would be a blunder from which there would be no return. I know of no technological means, no form of civil defense or counteroffensive warfare, which could reliably protect the biosphere from the catastrophic effects of a major nuclear war. There is, in my opinion, only one way to survive the threats of nuclear war—and that is to insure that it never happens. And because of the appreciable chance of an accidental nuclear war, I believe that the only way to do so is to destroy the world's stock of nuclear weapons and to develop less self-defeating means of protecting national security. Needless to say, the political difficulties involved in international nuclear disarmament are monumental.

Despite the dazzling successes of modern technology and the unprecedented power of modern military systems, they suffer from a common and catastrophic fault. While providing us with a bountiful supply of food, with great industrial plants, with high-speed transportation, and with military weapons of unprecedented power, they threaten our very survival. Technology has not only built the magnificent material base of modern society, but also confronts us with threats to survival which cannot be corrected unless we solve very grave economic, social, and political problems.

How can we explain this paradox? The answer is, I believe, that our technological society has committed a blunder familiar to us from the nineteenth century, when the dominant industries of the day, especially lumbering and mining, were successfully developed—by plundering the earth's natural resources. These industries provided cheap materials for constructing a new industrial society, but they accumulated a huge debt in destroyed and depleted resources, which had to be paid by later generations. The conservation movement was created in the United States to control these greedy assaults on our resources. The same thing is happening today, but now we are stealing from future generations not just their lumber or their coal, but the basic necessities of life: air, water, and soil. A new conservation movement is needed to preserve life itself.

The earlier ravages of our resources made very visible marks, but the new attacks are largely hidden. Thoughtless lumbering practices left vast scars on the land, but thoughtless development of modern industrial, agricultural, and military methods only gradually poison the air and water. Many of the pollutants—carbon dioxide, radioisotopes, pesticides, and excess nitrate—are invisible and go largely unnoticed until a lake dies, a river becomes foul, or children sicken. This time the world is being plundered in secret.

The earlier depredations on our resources were usually made with a fair knowledge of the harmful consequences, for it is difficult to escape the fact that erosion quickly follows the deforestation of a hillside. The difficulty lay not in scientific ignorance, but in willful greed. In the present situation, the hazards of modern pollutants are generally not appreciated until after the technologies which produce them are well established in the economy. While this ignorance absolves us from the immorality of the knowingly destructive acts that characterized the nineteenth century raids on our resources, the present fault is more serious. It signifies that the capability of science to guide us in our interventions into nature has been seriously eroded—that science has, indeed, got out of hand.

In this situation, scientists bear a very grave responsibility, for they are the guardians of the integrity of science. In the last few decades serious weaknesses in this system of principles have begun to appear. Secrecy has hampered free discourse. Major scientific enterprises have been governed by narrow national aims. In some cases, especially in the exploration of space, scientists have become so closely tied to basically political aims as to relinquish their traditional devotion to open discussion of conflicting views on what are often doubtful scientific conclusions.

What can scientists do to restore the integrity of science and to provide the kind of careful guidance to technology that is essential if we are to avoid catastrophic mistakes? No new principles are needed; instead, scientists need to find new ways to protect science itself from the encroachment of political pressures. This is not a new problem, for science and scholarship have often been under assault when their freedom to seek and to discuss the truth becomes a threat to existing economic or political power. The internal strength of science and its capability to understand nature have been weakened whenever the principles of scientific discourse were compromised, and restored when these principles were defended. The medieval suppressions of natural science, the perversion of science by Nazi racial theories, Soviet restraints on theories of genetics, and the suppression by United States military secrecy of open discussion of the Starfish project, have all been paid for in the most costly coin—knowledge. The lesson of all these experiences is the same. If science is to perform its duty to society, which is to guide, by objective knowledge, human interactions with the rest of nature, its integrity must be defended. Scientists must find ways to remove the restraints of secrecy, to insist on open discussion of the possible consequences of large-scale experiments *before* they are undertaken, to resist the hasty and unconditional support of conclusions that conform to the demands of current political or economic policy.

Apart from these duties toward science, I believe that scientists have a responsibility in relation to the technological uses which are made of scientific developments. In my opinion, the proper duty of the scientist to the social consequence of his work cannot be fulfilled by aloofness or by an approach which arrogates to scientists alone the social and moral judgments which are the right of every citizen. I propose that scientists are now bound by a new duty which adds to and extends their older responsibility for scholarship and teaching. We have the duty to inform, and to inform in keeping with the traditional principles of science, taking into account all relevant data and interpretations. This is an involuntary obligation to society; we have no right to withhold information from our fellow citizens, or to color its meaning with our own social judgments.

Scientists alone cannot accomplish these aims, for despite its tradition of independent scholarship, science is a dependent segment of society. In this sense defense of the integrity of science is a task for every citizen. And in this sense, too, the fate of science as a system of objective inquiry, and therefore its ability safely to guide the life of man on earth, will be determined by social intent. Both awareness of the grave social issues generated by new scientific knowledge, and the policy choices which these issues require, therefore become matters of public morality. Public morality will determine whether scientific inquiry remains free. Public morality will determine at what cost we shall enjoy freedom from insect pests, the convenience of automobiles, or the high productivity of agriculture. Only public morality can determine whether we ought to intrust our national security to the catastrophic potential of nuclear war.

There is a unique relationship between the scientist's social responsibilities and the general duties of citizenship. If the scientist, directly or by inferences from his actions, lays claim to a special responsibility for the resolution of the policy issues which relate to technology, he may, in effect, prevent others from performing their own political duties. If the scientist fails in his duty to inform citizens, they are precluded from the gravest acts of citizenship and lose their right of conscience.

We have been accustomed, in the past, especially in our organized systems of morality—religion—to exemplify the principles of moral life in terms which relate to Egypt under the pharaohs or Rome under the emperors. Since the establishment of Western religions, their custodians have, of course, labored to achieve a relevance to the changing states of society. In recent times the gap between traditional moral principles and the realities of modern life has become so large as to precipitate, beginning in the Catholic church, and less spectacularly in other religious denominations, urgent demands for renewal—for the development of statements of moral purpose which are directly relevant to the modern world. But in the modern world the substance of moral issues cannot be perceived in terms of the casting of stones or the theft of a neighbor's ox. The moral issues of the modern world are embedded in the complex substance of science and technology. The exercise of morality now requires the determination of right between the farmers whose pesticides poison the water and the fishermen whose livelihood may thereby be destroyed. It calls for a judgment between the advantages of replacing a smoky urban power generator with a smoke-free nuclear one which carries with it some hazard of a catastrophic accident. The ethical principles involved are no different from those invoked in earlier times, but the moral issues cannot be discerned unless the new substance in which they are expressed is understood. And since the substance of science is still often poorly perceived by most citizens, the technical content of the issues of the modern world shields them from moral judgment.

Nowhere is this more evident than in the case of nuclear war. The horrible face of nuclear war can only be described in scientific terms. It can be pictured only in the language of roentgens and megatonnage; it can be understood only by those who have some appreciation of industial organization, of human biology, of the intricacies of world-wide ecology. The self-destructiveness of nuclear war lies hidden behind a mask of science and technology. It is this shield, I believe, which has protected this most fateful moral issue in the history of man from the judgment of human morality. The greatest moral crime of our time is the concealment of the nature of nuclear war, for it deprives humanity of the solemn right to sit in judgment on its own fate; it condemns us all, unwittingly, to the greatest dereliction of conscience.

The obligation which our technological society forces upon all of us, scientist and citizen alike, is to discover how humanity can survive the new power which science has given it. It is already clear that even our present difficulties demand far-reaching social and political actions. Solution of our pollution problems will drastically affect the economic structure of the automobile industry, the power industry, and agriculture and will require basic changes in urban organization. To remove the threat of nuclear catastrophe we will be forced at last to resolve the pervasive international conflicts that have bloodied nearly every generation with war.

Every major advance in the technological competence of man has enforced revolutionary changes in the economic and political structure of society. The present age of technology is no exception to this rule of history. We already know the enormous benefits it can bestow; we have begun to perceive its frightful threats. The political crisis generated by this knowledge is upon us.

Science can reveal the depth of this crisis, but only social action can resolve it. Science can now serve society by exposing the crisis of modern technology to the judgment of all mankind. Only this judgment can determine whether the knowledge that science has given us shall destroy humanity or advance the welfare of man.

Rachel Carson (1907-1964) is surely the major prophet of the recent environmental movement. Although others had previously spoken out against the despoiling of the earth, it was her 1962 best seller Silent Spring *that informed the public of the dangers posed by the indiscriminate use of chemical poisons. In this selection, the final chapter from* Silent Spring, *the careful research and quiet articulateness which made her warnings so impressive are evident. She recommends biological controls as alternatives to the kinds of chemical overkill which she has documented, and which are also described by Paul Ehrlich, George Woodwell, and Frank Graham, Jr., in earlier essays in this volume. Like the oral contraceptive in the 1960's, new chemical pesticides and herbicides were considered a dazzling success upon their introduction in the 1940's; and, like the pill, these chemical poisons have proved to be something less than the perfect solution.*

Despite charges from the petrochemical industry and, to some extent, from the U. S. Department of Agriculture that her claims were alarmist and exaggerated, Rachel Carson has been vindicated by recent history. Unfortunately, she did not live to see such results of her labors as the recent imposition of bans upon the use of DDT and other chlorinated hydrocarbons. Miss Carson was a professional biologist and the author of the award-winning The Sea Around Us.

THE OTHER ROAD

by Rachel Carson

We stand now where two roads diverge. But unlike the roads in Robert Frost's familiar poem, they are not equally fair. The road we have long been traveling is deceptively easy, a smooth superhighway on which we progress with great speed, but at its end lies disaster. The other fork of the road—the one "less traveled by"—offers our last, our only chance to reach a destination that assures the preservation of our earth.

The choice, after all, is ours to make. If, having endured much, we have at last asserted our "right to know," and if, knowing, we have concluded that we are being asked to take senseless and frightening risks, then we should no longer accept the counsel of those who tell us that we must fill our world with poisonous chemicals; we should look about and see what other course is open to us.

A truly extraordinary variety of alternatives to the chemical control of insects is available. Some are already in use and have achieved brilliant success. Others are in the stage of laboratory testing. Still others are little

---

 In *Silent Spring*, this selection is accompanied by many, often highly specialized, bibliographical references. They are omitted here.

more than ideas in the minds of imaginative scientists, waiting for the opportunity to put them to the test. All have this in common: they are *biological* solutions, based on understanding of the living organisms they seek to control, and of the whole fabric of life to which these organisms belong. Specialists representing various areas of the vast field of biology are contributing—entomologists, pathologists, geneticists, physiologists, biochemists, ecologists—all pouring their knowledge and their creative inspirations into the formation of a new science of biotic controls.

"Any science may be likened to a river," says a Johns Hopkins biologist, Professor Carl P. Swanson. "It has its obscure and unpretentious beginning; its quiet stretches as well as its rapids; its periods of drought as well as of fullness. It gathers momentum with the work of many investigators and as it is fed by other streams of thought; it is deepened and broadened by the concepts and generalizations that are gradually evolved."

So it is with the science of biological control in its modern sense. In America it had its obscure beginnings a century ago with the first attempts to introduce natural enemies of insects that were proving troublesome to farmers, an effort that sometimes moved slowly or not at all, but now and again gathered speed and momentum under the impetus of an outstanding success. It had its period of drought when workers in applied entomology, dazzled by the spectacular new insecticides of the 1940's, turned their backs on all biological methods and set foot on "the treadmill of chemical control." But the goal of an insect-free world continued to recede. Now at last, as it has become apparent that the heedless and unrestrained use of chemicals is a greater menace to ourselves than to the targets, the river which is the science of biotic control flows again, fed by new streams of thought.

Some of the most fascinating of the new methods are those that seek to turn the strength of a species against itself—to use the drive of an insect's life forces to destroy it. The most spectacular of these approaches is the "male sterilization" technique developed by the chief of the United States Department of Agriculture's Entomology Research Branch, Dr. Edward Knipling, and his associates.

About a quarter of a century ago Dr. Knipling startled his colleagues by proposing a unique method of insect control. If it were possible to sterilize and release large numbers of insects, he theorized, the sterilized males would, under certain conditions, compete with the normal wild males so successfully that, after repeated releases, only infertile eggs would be produced and the population would die out.

The proposal was met with bureaucratic inertia and with skepticism from scientists, but the idea persisted in Dr. Knipling's mind. One major problem remained to be solved before it could be put to the test—a practical method of insect sterilization had to be found. Academically, the fact that insects could be sterilized by exposure to X-ray had been known since 1916, when an entomologist by the name of G. A. Runner reported such sterilization of cigarette beetles. Hermann Muller's pioneering work on the production of mutations by X-ray opened up vast new areas of thought in the late 1920's, and by the middle of the century various workers had reported the sterilization by X-rays or gamma rays of at least a dozen species of insects.

But these were laboratory experiments, still a long way from practical application. About 1950, Dr. Knipling launched a serious effort to turn insect sterilization into a weapon that would wipe out a major insect enemy of livestock in the South, the screw-worm fly. The females of this species lay their eggs in any open wound of a warm-blooded animal. The hatching larvae are parasitic, feeding on the flesh of the host. A full-grown steer may succumb to a heavy infestation in 10 days, and livestock losses in the United States have been estimated at $40,000,000 a year. The toll of wildlife is harder to measure, but it must be great. Scarcity of deer in some areas of Texas is attributed to the screw-worm. This is a tropical or subtropical insect, inhabiting South and Central America and Mexico, and in the United States normally restricted to the Southwest. About 1933, however, it was accidentally introduced into Florida, where the climate allowed it to survive over winter and to establish populations. It even pushed into southern Alabama and Georgia, and soon the livestock industry of the southeastern states was faced with annual losses running to $20,000,000.

A vast amount of information on the biology of the screw-worm had been accumulated over the years by Agriculture Department scientists in Texas. By 1954, after some preliminary field trials on Florida islands, Dr. Knipling was ready for a full-scale test of his theory. For this, by arrangement with the Dutch Government, he went to the island of Curaçao in the Caribbean, cut off from the mainland by at least 50 miles of sea.

Beginning in August 1954, screw-worms reared and sterilized in an Agriculture Department laboratory in Florida were flown to Curaçao and released from airplanes at the rate of about 400 per square mile per week. Almost at once the number of egg masses deposited on experimental goats began to decrease, as did their fertility. Only seven weeks after the releases were started, all eggs were infertile. Soon it was impossible to find a single egg mass, sterile or otherwise. The screw-worm had indeed been eradicated on Curaçao.

The resounding success of the Curaçao experiment whetted the appetites of Florida livestock raisers for a similar feat that would relieve them of the scourge of screw-worms. Although the difficulties here were relatively enormous—an area 300 times as large as the small Caribbean island—in 1957 the United States Department of Agriculture and the State of Florida joined in providing funds for an eradication effort. The project involved the weekly production of about 50 million screw-worms at a specially constructed "fly factory," the use of 20 light airplanes to fly pre-arranged flight patterns, five to six hours daily, each plane carrying a thousand paper cartons, each carton containing 200 to 400 irradiated flies.

The cold winter of 1957-58, when freezing temperatures gripped northern Florida, gave an unexpected opportunity to start the program while the screw-worm populations were reduced and confined to a small area. By the time the program was considered complete at the end of 17 months, 3 1/2 billion artificially reared, sterilized flies had been released over Florida and sections of Georgia and Alabama. The last-known animal wound infestation that could be attributed to screw-worm occurred in February 1959. In the next few weeks several adults were taken in traps. Thereafter no trace of the screw-worm could be discovered. Its extinction in the Southeast had been accomplished—a triumphant demonstration of the worth of scientific creativity, aided by thorough basic research, persistence, and determination.

Now a quarantine barrier in Mississippi seeks to prevent the re-entrance of the screw-worm from the Southwest, where it is firmly entrenched. Eradication there would be a formidable undertaking, considering the vast areas involved and the probability of re-invasion from Mexico. Nevertheless, the stakes are high and the thinking in the Department seems to be that some sort of program, designed at least to hold the screw-worm populations at very low levels, may soon be attempted in Texas and other infested areas of the Southwest.

The brilliant success of the screw-worm campaign has stimulated tremendous interest in applying the same methods to other insects. Not all, of course, are suitable subjects for this technique, much depending on details of the life history, population density, and reactions to radiation.

Experiments have been undertaken by the British in the hope that the method could be used against the tsetse fly in Rhodesia. This insect infests about a third of Africa, posing a menace to human health and preventing the keeping of livestock in an area of some 4 1/2 million square miles of wooded grasslands. The habits of the tsetse differ considerably from those of the screw-worm fly, and although it can be sterilized by radiation some technical difficulties remain to be worked out before the method can be applied.

The British have already tested a large number of other species for susceptibility to radiation. United States scientists have had some encouraging early results with the melon fly and the oriental and Mediterranean fruit flies in laboratory tests in Hawaii and field tests on the remote island of Rota. The corn borer and the sugarcane borer are also being tested. There are possibilities, too, that insects of medical importance might be controlled by sterilization. A Chilean scientist has pointed out that malaria-carrying mosquitoes persist in his country in spite of insecticide treatment; the release of sterile males might then provide the final blow needed to eliminate this population.

The obvious difficulties of sterilizing by radiation have led to search for an easier method of accomplishing similar results, and there is now a strongly running tide of interest in chemical sterilants.

Scientists at the Department of Agriculture laboratory in Orlando, Florida, are now sterilizing the housefly in laboratory experiments and even in some field trials, using chemicals incorporated in suitable foods. In a test on an island in the Florida Keys in 1961, a population of flies was nearly wiped out within a period of only five weeks. Repopulation of course followed from nearby islands, but as a pilot project the test was successful. The Department's excitement about the promise of this method is easily understood. In the first place, as we have seen, the housefly has now become virtually uncontrollable by insecticides. A completely new method of control is undoubtedly needed. One of the problems of sterilization by radiation is that this requires not only artificial rearing but the release of sterile males in larger number than are present in the wild population. This could be done with the screw-worm, which is actually not an abundant insect. With the housefly, however, more than doubling the population through releases could be highly objectionable, even though the increase would be only temporary. A chemical sterilant, on the other hand, could be combined with a bait substance and introduced into the natural environment of the fly; insects feeding on it would become sterile and in the course of time the sterile flies would predominate and the insects would breed themselves out of existence.

The testing of chemicals for a sterilizing effect is much more difficult than the testing of chemical poisons. It takes 30 days to evaluate one chemical —although, of course, a number of tests can be run concurrently. Yet between April 1958 and December 1961 several hundred chemicals were screened at the Orlando laboratory for a possible sterilizing effect. The Department of Agriculture seems happy to have found among these even a handful of chemicals that show promise.

Now other laboratories of the Department are taking up the problem, testing chemicals against stable flies, mosquitoes, boll weevils, and an assortment of fruit flies. All this is presently experimental but in the few years since work began on chemosterilants the project has grown enormously. In theory it has many attractive features. Dr. Knipling has pointed out that effective chemical insect sterilization "might easily outdo some of the best of known insecticides." Take an imaginary situation in which a population of a million insects is multiplying five times in each generation. An insecticide might kill 90 per cent of each generation, leaving 125,000 insects alive after the third generation. In contrast, a chemical that would produce 90 per cent sterility would leave only 125 insects alive.

On the other side of the coin is the fact that some extremely potent chemicals are involved. It is fortunate that at least during these early stages most of the men working with chemosterilants seem mindful of the need to find safe chemicals and safe methods of application. Nonetheless, suggestions are heard here and there that these sterilizing chemicals might be applied as aerial sprays—for example, to coat the foliage chewed by gypsy moth larvae. To attempt any such procedure without thorough advance research on the hazards involved would be the height of irresponsibility. If the potential hazards of the chemosterilants are not constantly borne in mind we could easily find ourselves in even worse trouble than that now created by the insecticides.

The sterilants currently being tested fall generally into two groups, both of which are extremely interesting in their mode of action. The first are intimately related to the life processes, or metabolism, of the cell; i.e., they so closely resemble a substance the cell or tissue needs that the organism "mistakes" them for the true metabolite and tries to incorporate them in its normal building processes. But the fit is wrong in some detail and the process comes to a halt. Such chemicals are called antimetabolites.

The second group consists of chemicals that act on the chromosomes, probably affecting the gene chemicals and causing the chromosomes to break up. The chemosterilants of this group are alkylating agents, which are extremely reactive chemicals, capable of intense cell destruction, damage to chromosomes, and production of mutations. It is the view of Dr. Peter Alexander of the Chester Beatty Research Institute in London that "any alkylating agent which is effective in sterilizing insects would also be a powerful mutagen and carcinogen." Dr. Alexander feels that any conceivable use of such chemicals in insect control would be "open to the most severe objections." It is to be hoped, therefore, that the present experiments will lead not to actual use of these particular chemicals but to the discovery of others that will be safe and also highly specific in their action on the target insect.

Some of the most interesting of the recent work is concerned with still other ways of forging weapons from the insect's own life processes. Insects produce a variety of venoms, attractants, repellants. What is the chemical nature of these secretions? Could we make use of them as, perhaps, very selective insecticides? Scientists at Cornell University and elsewhere are trying to find answers to some of these questions, studying the defense mechanisms by which many insects protect themselves from attack by predators, working out the chemical structure of insect secretions. Other scientists are working on the so-called "juvenile hormone," a powerful substance which prevents metamorphosis of the larval insect until the proper stage of growth has been reached.

Perhaps the most immediately useful result of this exploration of insect secretion is the development of lures, or attractants. Here again, nature has pointed the way. The gypsy moth is an especially intriguing example. The female moth is too heavy-bodied to fly. She lives on or near the ground, fluttering about among low vegetation or creeping up tree trunks. The male, on the contrary, is a strong flier and is attracted even from considerable distances by a scent released by the female from special glands. Entomologists have taken advantage of this fact for a good many years, laboriously preparing this sex attractant from the bodies of the female moths. It was then used in traps set for the males in census operations along the fringe of the insect's range. But this was an extremely expensive procedure.

Despite the much publicized infestations in the northeastern states, there were not enough gypsy moths to provide the material, and hand-collected female pupae had to be imported from Europe, sometimes at a cost of half a dollar per tip. It was a tremendous breakthrough, therefore, when, after years of effort, chemists of the Agriculture Department recently succeeded in isolating the attractant. Following upon this discovery was the successful preparation of a closely related synthetic material from a constituent of castor oil; this not only deceives the male moths but is apparently fully as attractive as the natural substance. As little as one microgram (1/1000, 000 gram) in a trap is an effective lure.

All this is of much more than academic interest, for the new and economical "gyplure" might be used not merely in census operations but in control work. Several of the more attractive possibilities are now being tested. In what might be termed an experiment in psychological warfare, the attractant is combined with a granular material and distributed by planes. The aim is to confuse the male moth and alter the normal behavior so that, in the welter of attractive scents, he cannot find the true scent trail leading to the female. This line of attack is being carried even further in experiments aimed at deceiving the male into attempting to mate with a spurious female. In the laboratory, male gypsy moths have attempted copulation with chips of wood, vermiculite, and other small, inanimate objects, so long as they were suitably impregnated with gyplure. Whether such diversion of the mating instinct into nonproductive channels would actually serve to reduce the population remains to be tested, but it is an interesting possibility.

The gypsy moth lure was the first insect sex attractant to be synthesized, but probably there will soon be others. A number of agricultural insects are being studied for possible attractants that man could imitate. Encouraging results have been obtained with the Hessian fly and the tobacco hornworm.

Combinations of attractants and poisons are being tried against several insect species. Government scientists have developed an attractant called methyl-eugenol, which males of the oriental fruit fly and the melon fly find irresistible. This has been combined with a poison in tests in the Bonin Islands 450 miles south of Japan. Small pieces of fiberboard were impregnated with the two chemicals and were distributed by air over the entire island chain to attract and kill the male flies. This program of "male annihilation" was begun in 1960: a year later the Agriculture Department estimated that more than 99 per cent of the population had been eliminated.

The method as here applied seems to have marked advantages over the conventional broadcasting of insecticides. The poison, an organic phosphorus chemical, is confined to squares of fiberboard which are unlikely to be eaten by wildlife; its residues, moreover, are quickly dissipated and so are not potential contaminants of soil or water.

But not all communication in the insect world is by scents that lure or repel. Sound also may be a warning or an attraction. The constant stream of ultrasonic sound that issues from a bat in flight (serving as a radar system to guide it through darkness) is heard by certain moths, enabling them to avoid capture. The wing sounds of approaching parasitic flies warn the larvae of some sawflies to herd together for protection. On the other hand, the sounds made by certain wood-boring insects enable their parasites to find them, and to the male mosquito the wing-beat of the female is a siren song.

What use, if any, can be made of this ability of the insect to detect and react to sound? As yet in the experimental stage, but nonetheless interesting, is the initial success in attracting male mosquitoes to playback recordings of the flight sound of the female. The males were lured to a charged grid and so killed. The repellant effect of bursts of ultrasonic sound is being tested in Canada against corn borer and cutworm moths. Two authorities on animal sound, Professors Hubert and Mable Frings of the University of Hawaii, believe that a field method of influencing the behavior of insects with sound only awaits discovery of the proper key to unlock and apply the vast existing knowledge of insect sound production and reception. Repellant sounds may offer greater possibilities than attractants. The Fringses are known for their discovery that starlings scatter in alarm before a recording of the distress cry of one of their fellows; perhaps somewhere in this fact is a central truth that may be applied to insects. To practical men of industry the possibilities seem real enough so that at least one major electronic corporation is preparing to set up a laboratory to test them.

Sound is also being tested as an agent of direct destruction. Ultrasonic sound will kill all mosquito larvae in a laboratory tank; however, it kills other aquatic organisms as well. In other experiments, blowflies, mealworms, and yellow fever mosquitoes have been killed by airborne ultrasonic sound in a matter of seconds. All such experiments are first steps toward wholly new concepts of insect control which the miracles of electronics may some day make a reality.

The new biotic control of insects is not wholly a matter of electronics and gamma radiation and other products of man's inventive mind. Some of its methods have ancient roots, based on the knowledge that, like ourselves, insects are subject to disease. Bacterial infections sweep through their populations like the plagues of old; under the onset of a virus their hordes sicken and die. The occurrence of disease in insects was known before the time of Aristotle; the maladies of the silkworm were celebrated in medieval poetry; and through study of the diseases of this same insect the first understanding of the principles of infectious disease came to Pasteur.

Insects are beset not only by viruses and bacteria but also by fungi, protozoa, microscopic worms, and other beings from all that unseen world of minute life that, by and large, befriends mankind. For the microbes include not only disease organisms but those that destroy waste matter, make soils fertile, and enter into countless biological processes like fermentation and nitrification. Why should they not also aid us in the control of insects?

One of the first to envision such use of microorganisms was the 19th-century zoologist Elie Metchnikoff. During the concluding decades of the 19th and the first half of the 20th centuries the idea of microbial control was slowly taking form. The first conclusive proof that an insect could be brought under control by introducing a disease into its environment came in the late 1930's with the discovery and use of milky disease for the Japanese beetle, which is caused by the spores of a bacterium belonging to the genus *Bacillus*. This classic example of bacterial control has a long history of use in the eastern part of the United States. . . .

High hopes now attend tests of another bacterium of this genus—*Bacillus thuringiensis*—originally discovered in Germany in 1911 in the province of Thuringia, where it was found to cause a fatal septicemia in the larvae of the flour moth. This bacterium actually kills by poisoning rather than by disease. Within its vegetative rods there are formed, along with spores, peculiar crystals composed of a protein substance highly toxic to certain insects, especially to the larvae of the mothlike lepidopteras. Shortly after eating foliage coated with this toxin the larva suffers paralysis, stops feeding, and soon dies. For practical purposes, the fact that feeding is interrupted promptly is of course an enormous advantage, for crop damage stops almost as soon as the pathogen is applied. Compounds containing spores of *Bacillus thuringiensis* are now being manufactured by several firms in the United States under various trade names. Field tests are being made in several countries: in France and Germany against larvae of the cabbage butterfly, in Yugoslavia against the fall webworm, in the Soviet Union against a tent caterpillar. In Panama, where tests were begun in 1961, this bacterial insecticide may be the answer to one or more of the serious problems confronting banana growers. There the root borer is a serious pest of the banana, so weakening its roots that the trees are easily toppled by wind. Dieldrin has been the only chemical effective against the borer, but it has now set in motion a chain of disaster. The borers are becoming resistant. The chemical has also destroyed some important insect predators and so has caused an increase in the tortricids—small, stout-bodied moths whose larvae scar the surface of the bananas. There is reason to hope the new microbial insecticide will eliminate both the tortricids and the borers and that it will do so without upsetting natural controls.

In eastern forests of Canada and the United States bacterial insecticides may be one important answer to the problems of such forest insects as the budworms and the gypsy moth. In 1960 both countries began field tests with a commercial preparation of *Bacillus thuringiensis.* Some of the early results have been encouraging. In Vermont, for example, the end results of bacterial control were as good as those obtained with DDT. The main technical problem now is to find a carrying solution that will stick the bacterial spores to the needles of the evergreens. On crops this is not a problem—even a dust can be used. Bacterial insecticides have already been tried on a wide variety of vegetables, especially in California.

Meanwhile, other perhaps less spectacular work is concerned with viruses. Here and there in California fields of young alfalfa are being sprayed with a substance as deadly as any insecticide for the destructive alfalfa caterpillar—a solution containing a virus obtained from the bodies of caterpillars that have died because of infection with this exceedingly virulent disease. The bodies of only five diseased caterpillars provide enough virus to treat an acre of alfalfa. In some Canadian forests a virus that affects pine sawflies has proved so effective in control that it has replaced insecticides.

Scientists in Czechoslovakia are experimenting with protozoa against webworms and other insect pests, and in the United States a protozoan parasite has been found to reduce the egg-laying potential of the corn borer.

To some the term microbial insecticide may conjure up pictures of bacterial warfare that would endanger other forms of life. This is not true. In contrast to chemicals, insect pathogens are harmless to all but their intended targets. Dr. Edward Steinhaus, an outstanding authority on insect pathology, has stated emphatically that there is "no authenticated recorded instance of a true insect pathogen having caused an infectious disease in a vertebrate animal either experimentally or in nature." The insect pathogens are so specific that they infect only a small group of insects—sometimes a single species. Biologically they do not belong to the type of organisms that cause disease in higher animals or in plants. Also, as Dr. Steinhaus points out, outbreaks of insect disease in nature always remain confined to insects, affecting neither the host plants nor animals feeding on them.

Insects have many natural enemies—not only microbes of many kinds but other insects. The first suggestion that an insect might be controlled by encouraging its enemies is generally credited to Erasmus Darwin about 1800. Probably because it was the first generally practiced method of biological control, this setting of one insect against another is widely but erroneously thought to be the only alternative to chemicals.

In the United States the true beginnings of conventional biological control date from 1888 when Albert Koebele, the first of a growing army of entomologist explorers, went to Australia to search for natural enemies of the cottony cushion scale that threatened the California citrus industry with destruction. . . . the mission was crowned with spectacular success, and in the century that followed the world has been combed for natural enemies to control the insects that have come uninvited to our shores. In all, about 100 species of imported predators and parasites have become established. Besides the vedalia beetles brought in by Koebele, other importations have been highly successful. A wasp imported from Japan established complete control of an insect attacking eastern apple orchards. Several natural enemies of the spotted alfalfa aphid, an accidental import from the Middle East, are credited with saving the California alfalfa industry. Parasites and predators of the gypsy moth achieved good control, as did the *Tiphia* wasp against the Japanese beetle. Biological control of scales and mealy bugs is estimated to save California several millions of dollars a year—indeed, one of the leading entomologists of that state, Dr. Paul DeBach, has estimated that for an investment of $4,000,000 in biological control work California has received a return of $100,000,000.

Examples of successful biological control of serious pests by importing their natural enemies are to be found in some 40 countries distributed over much of the world. The advantages of such control over chemicals are obvious: it is relatively inexpensive, it is permanent, it leaves no poisonous residues. Yet biological control has suffered from lack of support. California is virtually alone among the states in having a formal program in biological control, and many states have not even one entomologist who devotes full time to it. Perhaps for want of support biological control through insect enemies has not always been carried out with the scientific thoroughness it requires—exacting studies of its impact on the populations of insect prey have seldom been made, and releases have not always been made with the precision that might spell the difference between success and failure.

The predator and the preyed upon exist not alone, but as part of a vast web of life, all of which needs to be taken into account. Perhaps the opportunities for the more conventional types of biological control are greatest in the forests. The farmlands of modern agriculture are highly artificial, unlike anything nature ever conceived. But the forests are a different world, much closer to natural environments. Here, with a minimum of help and a maximum of noninterference from man, Nature can have her way, setting up all that wonderful and intricate system of checks and balances that protects the forest from undue damage by insects.

In the United States our foresters seem to have thought of biological control chiefly in terms of introducing insect parasites and predators. The Canadians take a broader view, and some of the Europeans have gone farthest of all to develop the science of "forest hygiene" to an amazing extent. Birds, ants, forest spiders, and soil bacteria are as much a part of a forest as the trees, in the view of European foresters, who take care to inoculate a new forest with these protective factors. The encouragement of birds is one of the first steps. In the modern era of intensive forestry the old hollow trees are gone and with them homes for woodpeckers and other tree-nesting birds. This lack is met by nesting boxes, which draw the birds back into the forest. Other boxes are specially designed for owls and for bats, so that these creatures may take over in the dark hours the work of insect hunting performed in daylight by the small birds.

But this is only the beginning. Some of the most fascinating control work in European forests employs the forest red ant as an aggressive insect predator—a species which, unfortunately, does not occur in North America. About 25 years ago Professor Karl Gösswald of the University of Würzburg developed a method of cultivating this ant and establishing colonies. Under his direction more than 10,000 colonies of the red ant have been established in about 90 test areas in the German Federal Republic.

Dr. Gösswald's method has been adopted in Italy and other countries, where ant farms have been established to supply colonies for distribution in the forests. In the Apennines, for example, several hundred nests have been set out to protect reforested areas.

"Where you can obtain in your forest a combination of birds' and ants' protection together with some bats and owls, the biological equilibrium has already been essentially improved," says Dr. Heinz Ruppertshofen, a forestry officer in Mölln, Germany, who believes that a single introduced predator or parasite is less effective than an array of the "natural companions" of the trees.

New ant colonies in the forests at Mölln are protected from woodpeckers by wire netting to reduce the toll. In this way the woodpeckers, which have increased by 400 per cent in 10 years in some of the test areas, do not seriously reduce the ant colonies, and pay handsomely for what they take by picking harmful caterpillars off the trees. Much of the work of caring for the ant colonies (and the birds' nesting boxes as well) is assumed by a youth corps from the local school, children 10 to 14 years old. The costs are exceedingly low; the benefits amount to permanent protection of the forests.

Another extremely interesting feature of Dr. Ruppertshofen's work is his use of spiders, in which he appears to be a pioneer. Although there is a large literature on the classification and natural history of spiders, it is scattered and fragmentary and deals not at all with their value as an agent of biological control. Of the 22,000 known kinds of spiders, 760 are native to Germany (and about 2000 to the United States). Twenty-nine families of spiders inhabit German forests.

To a forester the most important fact about a spider is the kind of net it builds. The wheel-net spiders are most important, for the webs of some of them are so narrow-meshed that they can catch all flying insects. A large web (up to 16 inches in diameter) of the cross spider bears some 120,000 adhesive nodules on its strands. A single spider may destroy in her life of 18 months an average of 2000 insects. A biologically sound forest has 50 to 150 spiders to the square meter (a little more than a square yard). Where there are fewer, the deficiency may be remedied by collecting and distributing the baglike cocoons containing the eggs. "Three cocoons of the wasp spider [which occurs also in America] yield a thousand spiders, which can catch 200,000 flying insects," says Dr. Ruppertshofen. The tiny and delicate young of the wheel-net spiders that emerge in the spring are especially important, he says, "as they spin in a teamwork a net umbrella above the top shoots of the trees and thus protect the young shoots against the flying insects." As the spiders molt and grow, the net is enlarged.

Canadian biologists have pursued rather similar lines of investigation, although with differences dictated by the fact that North American forests are largely natural rather than planted, and that the species available as aids in maintaining a healthy forest are somewhat different. The emphasis in Canada is on small mammals, which are amazingly effective in the control of certain insects, especially those that live within the spongy soil of the forest floor. Among such insects are the sawflies, so-called because the female has a saw-shaped ovipositor with which she slits open the needles of evergreen trees in order to deposit her eggs. The larvae eventually drop to the ground and form cocoons in the peat of tamarack bogs or the duff under spruce or pines. But beneath the forest floor is a world honeycombed with the tunnels and runways of small mammals—whitefooted mice, voles, and shrews of various species. Of all these small burrowers, the voracious shrews find and consume the largest number of sawfly cocoons. They feed by placing a forefoot on the cocoon and biting off the end, showing an extraordinary ability to discriminate between sound and empty cocoons. And for their insatiable appetite the shrews have no rivals. Whereas a vole can consume about 200 cocoons a day, a shrew, depending on the species, may devour up to 800! This may result, according to laboratory tests, in destruction of 75 to 98 per cent of the cocoons present.

It is not surprising that the island of Newfoundland, which has no native shrews but is beset with sawflies, so eagerly desired some of these small, efficient mammals that in 1958 the introduction of the masked shrew—the most efficient sawfly predator—was attempted. Canadian officials report in 1962 that the attempt has been successful. The shrews are multiplying and are spreading out over the island, some marked individuals having been recovered as much as ten miles from the point of release.

There is, then, a whole battery of armaments available to the forester who is willing to look for permanent solutions that preserve and strengthen the natural relations in the forest. Chemical pest control in the forest is at best a stopgap measure bringing no real solution, at worst killing the fishes in the forest streams, bringing on plagues of insects, and destroying the natural controls and those we may be trying to introduce. By such violent measures, says Dr. Ruppertshofen, "the partnership for life of the forest is entirely being unbalanced, and the catastrophes caused by parasites repeat in shorter and shorter periods . . . We, therefore, have to put an end to these unnatural manipulations brought into the most important and almost last natural living space which has been left for us."

Through all these new, imaginative, and creative approaches to the problem of sharing our earth with other creatures there runs a constant theme, the awareness that we are dealing with life—with living populations and all their pressures and counter-pressures, their surges and recessions. Only by taking account of such life forces and by cautiously seeking to guide them into channels favorable to ourselves can we hope to achieve a reasonable accommodation between the insect hordes and ourselves.

The current vogue for poisons has failed utterly to take into account these most fundamental considerations. As crude a weapon as the cave man's club, the chemical barrage has been hurled against the fabric of life—a fabric on the one hand delicate and destructible, on the other miraculously tough and resilient, and capable of striking back in unexpected ways. These extraordinary capacities of life have been ignored by the practitioners of chemical control who have brought to their task no "high-minded orientation," no humility before the vast forces with which they tamper.

The "control of nature" is a phrase conceived in arrogance, born of the Neanderthal age of biology and philosophy, when it was supposed that nature exists for the convenience of man. The concepts and practices of applied entomology for the most part date from that Stone Age of science. It is our alarming misfortune that so primitive a science has armed itself with the most modern and terrible weapons, and that in turning them against the insects it has also turned them against the earth.

In "To Survive on the Earth" Barry Commoner called for judgments to be made about the long-range environmental effects of new technological programs before they are begun. William O. Douglas (1898-   ) argues for a Cabinet-level Office of Conservation, staffed by professional biologists, which would be an authoritative voice for advising the President on programs involving the quality of our land, air, water, and wildlife. According to Douglas, there is no top-level conservation group which balances well-muscled and nearly autonomous federal organizations like the Corps of Engineers and the Bureau of Public Roads whose programs often result in the despoiling of our natural environment. Since Douglas wrote this in 1965, the idea of establishing an Office of Conservation or its equivalent has been increasingly heard and may soon become a reality.

William O. Douglas is a Justice of the Supreme Court and an inveterate outdoorsman, world traveler, and author.

OFFICE OF CONSERVATION

by William O. Douglas

At the federal level alone we have a vast medley of voices that make policies affecting conservation.

The Corps of Engineers—efficient, compact, politically well-organized—in many ways dominates the federal scene. There is hardly a river that is not embraced in its long-range plans for dams. It has a high degree of autonomy and aligns itself with pressure groups that make any proposed project an imminent threat. It has a preoccupation with dam building, though there is no reason why it could not be equally preoccupied with the design and construction of sewage disposal plants, of desalination of water, of the use of distillation or other processes to remove contaminants and nutrients from sewage effluent.

Other federal agencies also have an important say over our water resources. The Reclamation Bureau is one, and the Federal Power Commission, another. The former is concerned with building of dams for irrigation purposes; the latter, with building of dams for hydroelectric power. Those three plus existing agencies that administer waterway projects (TVA, Bonneville Authority, and the like) put the great weight of federal authority on the side of harnessing our rivers, of building more and still more dams, and of marking the demise of our free-flowing rivers.

Soon the Atomic Energy Commission will also be astride the water problem as nuclear energy becomes available to make sweet water out of salt water. Where its plants will be located and how their refuse will be disposed of are vital questions; all we know is that AEC decisions in the past have reflected more engineering than conservation values.

We have no voice of White House stature speaking on the broader sides of these vital issues.

When it comes to land use, the National Park Service, the United States Forest Service and BLM speak with authority in their respective domains. Yet the Bureau of Public Lands is an agency that has an overriding influence that is often destructive of the conservation values represented by those other federal agencies. The two federal agencies more destructive of wilderness values than any others are the Bureau of Public Roads and the Corps of Engineers. Each is efficient; each is politically minded; each is politically organized; each is very powerful in the nation's capital. These Twin Gods can bulldoze, pave, and dam the country until most of our sacred precincts are gone. The Corps of Engineers is obsessed with building dams across most of our navigable and non-navigable rivers. They have scouted everywhere and have blueprints that will ruin most of our free-flowing rivers. For the Middle Fork of the Salmon they propose nineteen dams. That bit of heaven on earth should be guarded against all invasions and preserved for all time. But the bait of vast sums of money is attractive; politicians like to bring contracts to their districts; and so the momentum is on the side of the Corps and against the wilderness.

Moreover, the work of the Corps is almost sacrosanct in Washington, D.C., for it has so many alliances that few ever speak out against it. The reason relates to the management of its budget. It contracts out much of its research to numerous federal agencies. A bureaucrat who can get funds from the Corps has it easy, for he need not run the gantlet of appropriation committees in Congress. It is administrative practice to take from 10 per cent to 15 per cent of a budget for overhead. So a Corps grant of $100,000 means $10,000 to $15,000 for building up the permanent staff. Parkinson's Law works well in this country as elsewhere; so these Corps grants are eagerly sought. Once acquired, their renewal is desired. Renewal cannot be expected if the alliance with the Corps is not a happy one. So efforts are made to generate goodwill. The result is that no important expert voice is ever raised against the Corps in Washington circles. It is the "white cow" above criticism and without fault.

The Bureau of Public Roads, like the Corps, has a large degree of autonomy. Its plans implicate local political organizations, each of which has its contractors. A road building program for an area such as Appalachia promises to feed local political machines when the planning should be at a higher level. Through highways can be built across Appalachia that will carry people at fast speeds through the area. But what is good for the soul, heart, and stomach of Appalachia? Modest roads and trails, rather than superhighways, are needed. Facilities for the enjoyment of the great wonders of Appalachia are needed, not turnpikes that treat Appalachia as a gateway to distant domains. The free-flowing rivers of Appalachia need protection and preservation, not destruction. Power plants are needed. But energy other than hydroelectric power must be used, saving the free-flowing rivers for the twenty-first century.

A powerful voice is needed to turn any federal agency from an objective it has chosen. Only the most powerful of all voices can stop the Bureau of Public Roads, the Corps of Engineers, or TVA from their programs for despoiling America's natural wonders.

Many issues of vital concern to naturalists and conservationists rest in the uncontrolled discretion of a bureaucracy. TVA decides because of low costs that Appalachia should be dug up and ruined forever in the cause of strip mining. BLM in spite of public outcry continues to fence the public out of public lands, making hunters and fishermen proceed at their peril, not knowing when they are on public or private lands. The Bureau of Public Roads under the hammer of chambers of commerce responds to state pressures to put a highway clear through a national forest. The Forest Service or BLM inaugurates spraying with pesticides, and the willow, on which the beavers exist and on which in turn the fish are dependent, is doomed.

Our federal policies are frequently conflicting.

We spent millions acquiring the Everglades—a swamp-type national park in Florida whose very existence depends on the flow of sweet water. We now spend more millions in drawing down the sweet water of the Everglades for other interests, jeopardizing the very existence of the park. We have given farmers federal assistance in draining wetlands so that they can grow crops we do not need. Those same wetlands are sorely needed as potholes for reproduction of ducks, whose population has been on the decline.

The examples could be numerous. The need is for an authoritative voice in the executive offices to call the attention of the President to these recurring crises so that he can determine what policy will in a given case be followed

—should a free-running river be sacrificed for a dam—should a spraying program be continued—should distillation of sewage effluent be our choice?

The President has the Office of Science and Technology, which has assisted greatly in matters pertaining to science. An Office of Conservation is also needed.

The main office of oversight in the conservation field which we have at present is the Bureau of Outdoor Recreation in the Interior Department, an office created by the Act of May 23, 1963. Edward C. Crafts is the Director and his superior is the Secretary of the Interior. The office maintains a continuing inventory and evaluation of outdoor recreation needs and resources; formulates a comprehensive nationwide recreation plan; recommends solutions whether by federal, state, or local government or by private interests; provides technical assistance and advice to the states and private interests with respect to outdoor recreation; sponsors and engages in research relating to outdoor recreation. To coordinate federal plans relating to outdoor recreation that office should have White House stature.

In 1949 England organized Nature Conservancy to create or perpetuate "nature reserves"—restriction of movement by people, vehicles, boats, and animals; restriction of hunting or fishing or the taking of eggs or interfering with the soil or the depositing of litter or lighting of fires; organizing research and scientific services.

It is a committee of the Privy Council and consists of not more than 18 nor less than 12 members. It administers some 110 "nature reserves," organizes and participates in conferences, proposes legislation, prepares educational materials, including films, on conservation, and conducts extensive scientific research on matters vital to wilderness values, including the use of toxic chemicals in agriculture. Its voice in affairs of state is already a potent one, which rallies public opinion quickly and summons official action at once.

We do not need an additional agency or office to manage areas like parks, forests, and seashores. But we do need one that gives the President an overriding value judgment on whether the sacrifice or loss of conservation values in this project or that proposal is worthy of our civilization.

The idea is not a new one. The late Senator Clair Engle of California and Senator Gale McGee of Wyoming introduced bills in 1961 to create such an office, one bill calling it the Resources and Conservation Council in the Executive Office of the President. Standards were made specific—a national purpose to promote "conditions under which there will be conservation, development, and utilization of the natural resources of the nation to meet human, economic, and national defense requirements, including recreational, wildlife, scenic, and scientific values and the enhancement of the national heritage for future generations."

There were hearings on these bills and the opposition was great. The Bureau of the Budget is staffed with people primarily interested in accounting and fiscal matters. The Office of Conservation should be trained in biology and all the allied skills, going into the preservation and propagation of wildlife and of the earth itself. An Office of Conservation is needed, not to make for better coordination of existing policies but to show what a projection of those policies for a decade or more means.

Water for irrigation—water for home and factory use—water for swimming and canoeing—how can it be attained?

Trails in valleys and along ridges where no mechanized vehicles are allowed—how can they be so plentiful as not to be crowded, come the twenty-first century?

Pesticides and herbicides, spraying and dusting of gardens, fields, and forests—which one should be banned, which put on probation, which tolerated?

Wooded sanctuaries free of roads, the ax, and the bulldozer—how can they be multiplied so that every child will have some chance for adventure into the unknown?

Rolling grasslands, lush alpine basins, primitive meadowlands—how can they be increased?

Bird and wildlife sanctuaries in ample number and variety so that no species becomes extinct—how can this be managed?

Picnic grounds for the autoist and nature trails for the pedestrian—how can these be multipled a thousandfold?

An Office of Conservation can help the President act in an informed way. It will give him the conservation values at stake in a particular decision. At present there is often a medley of voices when it comes to the public good and conservation needs. Presidents are human and they too make mistakes. But when it comes to the earth and our management of it, the costly mistakes of the past must be avoided. England has found the way in her Nature Conservancy. We can find ours in an Office of Conservation and perhaps become educated in man's civilized relation to the earth and its wildness.

Like many of the other authors presented in this book, Kenneth E. Boulding (1910-   ) is concerned with changing our traditional ways of thinking. An economist, Boulding explains the radical changes in accepted economic theory which will become necessary as Americans become aware that we are not cowboys riding limitless plains but spacemen aboard a closed spaceship—the earth. Nor is the spaceman economy to be discounted because it is still in the future; as the author says, this future is in many ways already upon us.

Boulding points out that recycling—turning production's waste products into raw materials for new production—will be a necessity aboard the closed spaceship earth. Although he does not pursue the subject, recycling is today receiving increasing attention. Garbage can be pressed into building blocks; aluminum cans, cardboard, and newspapers can be collected and recycled; vegetable wastes from the kitchen can be used for garden compost. These and other kinds of recycling can be expected to increase in the future as we become more aware of the earth as a spaceship.

Kenneth E. Boulding was born in Liverpool, England, and received his B.A. and M.A. from Oxford. He became a United States citizen in 1948 and is now a professor of economics.

THE ECONOMICS OF THE COMING SPACESHIP EARTH

by Kenneth E. Boulding

We are now in the middle of a long process of transition in the nature of the image which man has of himself and his environment. Primitive men, and to a large extent also men of the early civilizations, imagined themselves to be living on a virtually illimitable plane. There was almost always somewhere beyond the known limits of human habitation, and over a very large part of the time that man has been on earth, there has been something like a frontier. That is, there was always some place else to go when things got too difficult, either by reason of the deterioration of the natural environment or a deterioration of the social structure in places where people happened to live. The image of the frontier is probably one of the oldest images of mankind, and it is not surprising that we find it hard to get rid of.

Gradually, however, man has been accustoming himself to the notion of the spherical earth and a closed sphere of human activity. A few unusual spirits among the ancient Greeks perceived that the earth was a sphere. It was only with the circumnavigations and the geographical explorations of

---

"The Economics of the Coming Spaceship Earth" From *Environmental Quality in a Growing Economy*: Essays from the Sixth Resources for the Future Forum, edited by Henry Jarrett, published for Resources for the Future, Inc., by The Johns Hopkins Press.

the fifteenth and sixteenth centuries, however, that the fact that the earth was a sphere became at all widely known and accepted. Even in the nineteenth century, the commonest map was Mercator's projection, which visualizes the earth as an illimitable cylinder, essentially a plane wrapped around the globe, and it was not until the Second World War and the development of the air age that the global nature of the planet really entered the popular imagination. Even now we are very far from having made the moral, political, and psychological adjustments which are implied in this transition from the illimitable plane to the closed sphere.

Economists in particular, for the most part, have failed to come to grips with the ultimate consequences of the transition from the open to the closed earth. One hesitates to use the terms "open" and "closed" in this connection, as they have been used with so many different shades of meaning. Nevertheless, it is hard to find equivalents. The open system, indeed, has some similarities to the open system of von Bertalanffy,[1] in that it implies that some kind of a structure is maintained in the midst of a throughput from inputs to outputs. In a closed system, the outputs of all parts of the system are linked to the inputs of other parts. There are no inputs from outside and no outputs to the outside; indeed, there is no outside at all. Closed systems, in fact, are very rare in human experience, in fact almost by definition unknowable, for if there are genuinely closed systems around us, we have no way of getting information into them or out of them; and hence if they are really closed, we would be quite unaware of their existence. We can only find out about a closed system if we participate in it. Some isolated primitive societies may have approximated to this, but even these had to take inputs from the environment and give outputs to it. All living organisms, including man himself, are open systems. They have to receive inputs in the shape of air, food, water, and give off outputs in the form of effluvia and excrement. Deprivation of input of air, even for a few minutes, is fatal. Deprivation of the ability to obtain any input or to dispose of any output is fatal in a relatively short time. All human societies have likewise been open systems. They receive inputs from the earth, the atmosphere, and the water, and they give outputs into these reservoirs; they also produce inputs internally in the shape of babies and outputs in the shape of corpses. Given a capacity to draw upon inputs and to get rid of outputs, an open system of this kind can persist indefinitely.

There are some systems—such as the biological phenotype, for instance the human body—which cannot maintain themselves indefinitely by inputs and outputs because of the phenomenon of aging. This process is very little understood. It occurs, evidently, because there are some outputs which cannot be replaced by any known input. There is not the same necessity for aging in organizations and in societies, although an analogous phenomenon may take place. The structure and composition of an organization or society, however, can be maintained by inputs of fresh personnel from birth and education as the existing personnel ages and eventually dies. Here we have an interesting example of a system which seems to maintain itself by the self-generation of inputs, and in this sense is moving towards closure. The input of people (that is, babies) is also an output of people (that is, parents).

Systems may be open or closed in respect to a number of classes of inputs and outputs. Three important classes are matter, energy, and information. The present world economy is open in regard to all three. We can think of the world economy or "econosphere" as a subset of the "world set," which is the set of all objects of possible discourse in the world. We then think of the state of the econosphere at any one moment as being the total capital stock, that is, the set of all objects, people, organizations, and so on, which are interesting from the point of view of the system of exchange. This total stock of capital is clearly an open system in the sense that it has inputs and outputs, inputs being production which adds to the capital stock, outputs being consumption which subtracts from it. From a material point of view, we see objects passing from the noneconomic into the economic set in the process of production, and we similarly see products passing out of the economic set as their value becomes zero. Thus we see the econosphere as a material process involving the discovery and mining of fossil fuels, ores, etc., and at the other end a process by which the effluents of the system are passed out into noneconomic reservoirs—for instance, the atmosphere and the oceans—which are not appropriated and do not enter into the exchange system.

From the point of view of the energy system, the econosphere involves inputs of available energy in the form, say, of water power, fossil fuels, or sunlight, which are necessary in order to create the material throughput and to move matter from the noneconomic set into the economic set or even out of it again; and energy itself is given off by the system in a less available form, mostly in the form of heat. These inputs of available energy must come either from the sun (the energy supplied by other stars being assumed to be negligible) or it may come from the earth itself, either through its internal heat or through its energy of rotation or other motions, which generate, for instance, the energy of the tides. Agriculture, a few solar machines, and water power use the current available energy income. In advanced societies this is supplemented very extensively by the use of fossil fuels, which represent as it were a capital stock of stored-up sunshine. Because of this capital stock of energy, we have been able to maintain an energy input into the system, particularly over the last two centuries, much larger than we would have been able to do with existing techniques if we had had to rely on the current input of available energy from the sun or the earth itself. This supplementary input, however, is by its very nature exhaustible.

The inputs and outputs of information are more subtle and harder to trace, but also represent an open system, related to, but not wholly dependent on, the transformations of matter and energy. By far the larger amount of information and knowledge is self-generated by the human society, though a certain amount of information comes into the sociosphere in the form of light from the universe outside. The information that comes from the universe has certainly affected man's image of himself and of his environment, as we can easily visualize if we suppose that we lived on a planet with a total cloud-cover that kept out all information from the exterior universe. It is only in very recent times, of course, that the information coming in from the universe has been captured and coded into the form of a complex image of what the universe is like outside the earth; but even in primitive times, man's perception of the heavenly bodies has always profoundly affected his image of earth and of himself. It is the information generated within the planet, however, and particularly that generated by man himself, which forms by far the larger part of the information system. We can think of the stock of knowledge, or as Teilhard de Chardin called it, the "noosphere," and consider this as an open system, losing knowledge through aging and death and gaining it through birth and education and the ordinary experience of life.

From the human point of view, knowledge or information is by far the most important of the three systems. Matter only acquires significance and only enters the sociosphere or the econosphere insofar as it becomes an object of human knowledge. We can think of capital, indeed, as frozen knowledge or knowledge imposed on the material world in the form of improbable arrangements. A machine, for instance, originated in the mind of man, and both its construction and its use involve information processes imposed on the material world by man himself. The cumulation of knowledge, that is, the excess of its production over its consumption, is the key to human development of all kinds, especially to economic development. We can see this pre-eminence of knowledge very clearly in the experiences of countries where the material capital has been destroyed by a war, as in Japan and Germany. The knowledge of the people was not destroyed, and it did not take long, therefore, certainly not more than ten years, for most of the material capital to be reestablished again. In a country such as Indonesia, however, where the knowledge did not exist, the material capital did not come into being either. By "knowledge" here I mean, of course, the whole cognitive structure, which includes valuations and motivations as well as images of the factual world.

The concept of entropy, used in a somewhat loose sense, can be applied to all three of these open systems. In the case of material systems, we can distinguish between entropic processes, which take concentrated materials and diffuse them through the oceans or over the earth's surface or into the atmosphere, and anti-entropic processes, which take diffuse materials and concentrate them. Material entropy can be taken as a measure of the uniformity of the distribution of elements and, more uncertainly, compounds and other structures on the earth's surface. There is, fortunately, no law of increasing material entropy, as there is in the corresponding case of energy, as it is quite possible to concentrate diffused materials if energy inputs are allowed. Thus the processes for fixation of nitrogen from the air, processes for the extraction of magnesium or other elements from the sea, and processes for the desalinization of sea water are anti-entropic in the material sense, though the reduction of material entropy has to be paid for by inputs of energy and also inputs of information, or at least a stock of information in the system. In regard to matter, therefore, a closed system is conceivable, that is, a system in which there is neither increase nor decrease in material entropy. In such a system all outputs from consumption would constantly be recycled to become inputs for production, as for instance, nitrogen in the nitrogen cycle of the natural ecosystem.

In regard to the energy system there is, unfortunately, no escape from the grim Second Law of Thermodynamics; and if there were no energy inputs into the earth, any evolutionary or developmental process would be impossible. The large energy inputs which we have obtained from fossil fuels are strictly temporary. Even the most optimistic predictions would expect the easily available supply of fossil fuels to be exhausted in a mere matter of centuries at present rates of use. If the rest of the world were to rise to American standards of power consumption, and still more if world population continues to increase, the exhaustion of fossil fuels would be even more rapid. The development of nuclear energy has improved this picture, but has not fundamentally altered it, at least in present technologies, for fissionable material is still relatively scarce. If we should achieve the economic use of energy through fusion, of course, a much larger source of energy materials would be available, which would expand the time horizons of supplementary energy input into an open social system by perhaps tens to hundreds of thousands of years. Failing this, however, the time is not very far distant, historically speaking, when man will once more have to retreat to his current energy input from the sun, even though this could be used much more effectively than in the past with increased knowledge. Up to now, certainly, we have not got very far with the technology of using current solar energy, but the possibility of substantial improvements in the future is certainly high. It may be, indeed, that the biological revolution which is just beginning will produce a solution to this problem, as we develop artificial organisms which are capable of much more efficient transformation of solar energy into easily available forms than any that we now have. As Richard Meier has suggested, we may run our machines in the future with methane-producing algae.

The question of whether there is anything corresponding to entropy in the information system is a puzzling one, though of great interest. There are certainly many examples of social systems and cultures which have lost knowledge, especially in transition from one generation to the next, and in which the culture has therefore degenerated. One only has to look at the folk culture of Appalachian migrants to American cities to see a culture which started out as a fairly rich European folk culture in Elizabethan times and which seems to have lost both skills, adaptability, folk tales, songs, and almost everything that goes up to make richness and complexity in a culture, in the course of about ten generations. The American Indians on reservations provide another example of such degradation of the information and knowledge system. On the other hand, over a great part of human history, the growth of knowledge in the earth as a whole seems to have been almost continuous, even though there have been times of relatively slow growth and times of rapid growth. As it is knowledge of certain kinds that produces the growth of knowledge in general, we have here a very subtle and complicated system, and it is hard to put one's finger on the particular elements in a culture which make knowledge grow more or less rapidly, or even which make it decline. One of the great puzzles in this connection, for instance, is why the take-off into science, which represents an "acceleration," or an increase in the rate of growth of knowledge in

European society in the sixteenth century, did not take place in China, which at that time (about 1600) was unquestionably ahead of Europe, and one would think even more ready for the breakthrough. This is perhaps the most crucial question in the theory of social development, yet we must confess that it is very little understood. Perhaps the most significant factor in this connection is the existence of "slack" in the culture, which permits a divergence from established patterns and activity which is not merely devoted to reproducing the existing society but is devoted to changing it. China was perhaps too well-organized and had too little slack in its society to produce the kind of acceleration which we find in the somewhat poorer and less well-organized but more diverse societies of Europe.

The closed earth of the future requires economic principles which are somewhat different from those of the open earth of the past. For the sake of picturesqueness, I am tempted to call the open economy the "cowboy economy," the cowboy being symbolic of the illimitable plains and also associated with reckless, exploitative, romantic, and violent behavior, which is characteristic of open societies. The closed economy of the future might similarly be called the "spaceman" economy, in which the earth has become a single spaceship, without unlimited reservoirs of anything, either for extraction or for pollution, and in which, therefore, man must find his place in a cyclical ecological system which is capable of continuous reproduction of material form even though it cannot escape having inputs of energy. The difference between the two types of economy becomes most apparent in the attitude towards consumption. In the cowboy economy, consumption is regarded as a good thing and production likewise; and the success of the economy is measured by the amount of the throughput from the "factors of production," a part of which, at any rate, is extracted from the reservoirs of raw materials and noneconomic objects, and another part of which is output into the reservoirs of pollution. If there are infinite reservoirs from which material can be obtained and into which effluvia can be deposited, then the throughput is at least a plausible measure of the success of the economy. The gross national product is a rough measure of this total throughput. It should be possible, however, to distinguish that part of the GNP which is derived from exhaustible and that which is derived from reproducible resources, as well as that part of consumption which represents effluvia and that which represents input into the productive system again. Nobody, as far as I know, has ever attempted to break down the GNP in this way, although it would be an interesting and extremely important exercise, which is unfortunately beyond the scope of this paper.

By contrast, in the spaceman economy, throughput is by no means a desideratum, and is indeed to be regarded as something to be minimized rather than maximized. The essential measure of the success of the economy is not production and consumption at all, but the nature, extent, quality, and complexity of the total capital stock, including in this the state of the human bodies and minds included in the system. In the spaceman economy, what we are primarily concerned with is stock maintenance, and any technological change which results in the maintenance of a given total stock with a lessened throughput (that is, less production and consumption) is clearly a gain. This idea that both production and consumption are bad things rather than good things is very strange to economists, who have been obsessed with the income-flow concepts to the exclusion, almost, of capital-stock concepts.

There are actually some very tricky and unsolved problems involved in the questions as to whether human welfare or well-being is to be regarded as a stock or a flow. Something of both these elements seems actually to be involved in it, and as far as I know there have been practically no studies directed towards identifying these two dimensions of human satisfaction. Is it, for instance, eating that is a good thing, or is it being well fed? Does economic welfare involve having nice clothes, fine houses, good equipment, and so on, or is it to be measured by the depreciation and the wearing out of these things? I am inclined myself to regard the stock concept as most fundamental, that is, to think of being well fed as more important than eating, and to think even of so-called services as essentially involving the restoration of a depleting psychic capital. Thus I have argued that we go to a concert in order to restore a psychic condition which might be called "just having gone to a concert," which, once established, tends to depreciate. When it depreciates beyond a certain point, we go to another concert in order to restore it. If it depreciates rapidly, we go to a lot of concerts; if it depreciates slowly, we go to few. On this view, similarly, we eat primarily to restore bodily homeostasis, that is, to maintain a condition of being well fed, and so on. On this view, there is nothing desirable in consumption at all. The less consumption we can maintain a given state with, the better off we are. If we had clothes that did not wear out, houses that did not depreciate, and even if we could maintain our bodily condition without eating, we would clearly be much better off.

It is this last consideration, perhaps, which makes one pause. Would we, for instance, really want an operation that would enable us to restore all our bodily tissues by intravenous feeding while we slept? Is there not, that is to say, a certain virtue in throughput itself, in activity itself, in production and consumption itself, in raising food and in eating it? It would certainly be rash to exclude this possibility. Further interesting problems are raised by the demand for variety. We certainly do not want a constant state to be maintained; we want fluctuations in the state. Otherwise there would be no demand for variety in food, for variety in scene, as in travel, for variety in social contact, and so on. The demand for variety can, of course, be costly, and sometimes it seems to be too costly to be tolerated or at least legitimated, as in the case of marital partners, where the maintenance of a homeostatic state in the family is usually regarded as much more desirable than the variety and excessive throughput of the libertine. There are problems here which the economics profession has neglected with astonishing singlemindedness. My own attempts to call attention to some of them, for instance, in two articles, as far as I can judge, produced no response whatever; and economists continue to think and act as if production, consumption, throughput, and the GNP were the sufficient and adequate measure of economic success.

It may be said, of course, why worry about all this when the spaceman economy is still a good way off (at least beyond the lifetimes of any now living), so let us eat, drink, spend, extract and pollute, and be as merry as we can, and let posterity worry about the spaceship earth. It is always a little hard to find a convincing answer to the man who says, "What has posterity ever done for me?" and the conservationist has always had to fall back on rather vague ethical principles postulating identity of the individual with some human community or society which extends not only back into the past but forward into the future. Unless the individual identifies with some community of this kind, conservation is obviously "irrational." Why should we not maximize the welfare of this generation at the cost of posterity? "*Après nous, le déluge* has been the motto of not insignificant numbers of human societies. The only answer to this, as far as I can see, is to point out that the welfare of the individual depends on the extent to which he can identify himself with others, and that the most satisfactory individual identity is that which identifies not only with a community in space but also with a community extending over time from the past into the future. If this kind of identity is recognized as desirable, then posterity has a voice, even if it does not have a vote; and in a sense, if its voice can influence votes, it has votes too. This whole problem is linked up with the much larger one of the determinants of the morale, legitimacy, and "nerve" of a society, and there is a great deal of historical evidence to suggest that a society which loses its identity with posterity and which loses its positive image of the future loses also its capacity to deal with present problems, and soon falls apart.[4]

Even if we concede that posterity is relevant to our present problems, we still face the question of time-discounting and the closely related question of uncertainty-discounting. It is a well-known phenomenon that individuals discount the future, even in their own lives. The very existence of a positive rate of interest may be taken as at least strong supporting evidence of this hypothesis. If we discount our own future, it is certainly not unreasonable to discount posterity's future even more, even if we do give posterity a vote. If we discount this at 5 per cent per annum, posterity's vote or dollar halves every fourteen years as we look into the future, and after even a mere hundred years it is pretty small—only about 1 1/2 cents on the dollar. If we add another 5 per cent for uncertainty, even the vote of our grandchildren reduces almost to insignificance. We can argue, of course, that the ethical thing to do is not to discount the future at all, that time-discounting is mainly the result of myopia and perspective, and hence is an illusion which the moral man should not tolerate. It is a very popular illusion, however, and one that must certainly be taken into consideration in the formulation of policies. It explains, perhaps, why conservationist policies almost have to be sold under some other excuse which seems more urgent, and why, indeed, necessities which are visualized as urgent, such as defense, always seem to hold priority over those which involve the future.

All these considerations add some credence to the point of view which says that we should not worry about the spaceman economy at all, and that we should just go on increasing the GNP and indeed the gross world product, or GWP, in the expectation that the problems of the future can be left to the future, that when scarcities arise, whether this is of raw materials or of pollutable reservoirs, the needs of the then present will determine the solutions of the then present, and there is no use giving ourselves ulcers by worrying about problems that we really do not have to solve. There is even high ethical authority for this point of view in the New Testament, which advocates that we should take no thought for tomorrow and let the dead bury their dead. There has always been something rather refreshing in the view that we should live like the birds, and perhaps posterity is for the birds in more senses than one; so perhaps we should all call it a day and go out and pollute something cheerfully. As an old taker of thought for the morrow, however, I cannot quite accept this solution; and I would argue, furthermore, that tomorrow is not only very close, but in many respects it is already here. The shadow of the future spaceship, indeed, is already falling over our spendthrift merriment. Oddly enough, it seems to be in pollution rather than in exhaustion that the problem is first becoming salient. Los Angeles has run out of air, Lake Erie has become a cesspool, the oceans are getting full of lead and DDT, and the atmosphere may become man's major problem in another generation, at the rate at which we are filling it up with gunk. It is, of course, true that at least on a microscale, things have been worse at times in the past. The cities of today, with all their foul air and polluted waterways, are probably not as bad as the filthy cities of the pretechnical age. Nevertheless, that fouling of the nest which has been typical of man's activity in the past on a local scale now seems to be extending to the whole world society; and one certainly cannot view with equanimity the present rate of pollution of any of the natural reservoirs, whether the atmosphere, the lakes, or even the oceans.

I would argue strongly also that our obsession with production and consumption to the exclusion of the "state" aspects of human welfare distorts the process of technological change in a most undesirable way. We are all familiar, of course, with the wastes involved in planned obsolescence, in competitive advertising, and in poor quality of consumer goods. These problems may not be so important as the "view with alarm" school indicates, and indeed the evidence at many points is conflicting. New materials especially seem to edge towards the side of improved durability, such as, for instance, neolite soles for footwear, nylon socks, wash and wear shirts, and so on. The case of household equipment and automobiles is a little less clear. Housing and building construction generally almost certainly has declined in durability since the Middle Ages, but this decline also reflects a change in tastes towards flexibility and fashion and a need for novelty, so that it is not easy to assess. What is clear is that no serious attempt has been made to assess the impact over the whole of economic life of changes in durability, that is, in the ratio of capital in the widest possible sense to income. I suspect that we have underestimated, even in our spendthrift society, the gains from increased durability, and that this might very well be one of the places where the price system needs correction through government-sponsored research and development. The problems which the spaceship earth is going to present, therefore, are not all in the future by any means, and a strong case can be made for paying much more attention to them in the present than we now do.

It may be complained that the considerations I have been putting forth relate only to the very long run, and they do not much concern our immediate problems. There may be some justice in this criticism, and my main excuse is that other writers have dealt adequately with the more immediate problems of deterioration in the quality of the environment. It is true, for instance, that many of the immediate problems of pollution of the atmosphere or of bodies of water arise because of the failure of the price system, and many of them could be solved by corrective taxation. If people had to pay the losses due to the nuisances which they create, a good deal more resources would go into the prevention of nuisances. These arguments involving external economies and diseconomies are familiar to economists, and there is no need to recapitulate them. The law of torts is quite inadequate to provide for the correction of the price system which is required, simply because where damages are widespread and their incidence on any particular person is small, the ordinary remedies of the civil law are quite inadequate and inappropriate. There needs, therefore, to be special legislation to cover these cases, and though such legislation seems hard to get in practice, mainly because of the widespread and small personal incidence of the injuries, the technical problems involved are not insuperable. If we were to adopt in principle a law for tax penalties for social damages, with an apparatus for making assessments under it, a very large proportion of current pollution and deterioration of the environment would be prevented. There are tricky problems of equity involved, particularly where old established nuisances create a kind of "right by purchase" to perpetuate themselves, but these are problems again which a few rather arbitrary decisions can bring to some kind of solution.

The problems which I have been raising in this paper are of larger scale and perhaps much harder to solve than the more practical and immediate problems of the above paragraph. Our success in dealing with the larger problems, however, is not unrelated to the development of skill in the solution of the more immediate and perhaps less difficult problems. One can hope, therefore, that as a succession of mounting crises, especially in pollution, arouse public opinion and mobilize support for the solution of the immediate problems, a learning process will be set in motion which will eventually lead to an appreciation of and perhaps solutions for the larger ones. My neglect of the immediate problems, therefore, is in no way intended to deny their importance, for unless we at least make a beginning on a process for solving the immediate problems we will not have much chance of solving the larger ones. On the other hand, it may also be true that a long-run vision, as it were, of the deep crisis which faces mankind may predispose people to taking more interest in the immediate problems and to devote more effort for their solution. This may sound like a rather modest optimism, but perhaps a modest optimism is better than no optimism at all.

In this final article, Aldous Huxley (1894-1963) engages the central problem which plagues those who have come to know something of the importance of ecology to man's survival: How can ecological awareness develop into social and political action? Huxley recognizes that power is now largely in the hands of rulers committed to "bad and unrealistic" policies; that is, policies which ignore the ecological threats pressed upon us by the headlong rush of modern science and technology. He maintains that our survival depends upon focussing collective attention on the biological aspects of man's condition. Unlike the problems of power politics and nationalism which defy rational analysis and often lead to war, the problems of ecology can yield to rational analysis. By having governments deal with the problems of human ecology instead of practicing power politics, we may be insuring the future in two ways: The threat of sudden destruction by scientific war is reduced and the possibility of biological disaster, however gradual, is reduced. Although Huxley's essay was written in 1963, much of his argument is still valid.

Aldous Huxley, younger brother of Sir Julian Huxley, was an English novelist and essayist who lived in his later years in California. His lifelong concern about science, which he shared with his brother, is seen in his most famous novel Brave New World, *a savage satire on life in a science-created utopia.*

THE POLITICS OF POPULATION / SECOND EDITION

by Aldous Huxley

In politics, the central and fundamental problem is the problem of power. Who is to exercise power? And by what means, by what authority, with what purpose in view, and under what controls? Yes, under what controls? For, as history has made it abundantly clear, to possess power is *ipso facto* to be tempted to abuse it. In mere self-preservation we must create and maintain institutions that make it difficult for the powerful to be led into those temptations which, succumbed to, transform them into tyrants at home and imperialists abroad.

For this purpose what kind of institutions are effective? And, having created them, how then can we guarantee them against obsolescence? Circumstances change, and, as they change, the old, the once so admirably effective devices for controlling power cease to be adequate. What then? Specifically, when advancing science and acceleratingly progressive technology alter man's long-established relationship with the planet on which he lives, revolutionize his societies, and at the same time equip his rulers with new and immensely more powerful instruments of domination, what ought we to do? What *can* we do?

"The Politics of Population/Second Edition" Reprinted, by permission, from the March 1969 issue of *The Center Magazine*, a publication of the Center for the Study of Democratic Institutions in Santa Barbara, California. Originally published in 1963 by the Center as an Occasional Paper, "The Politics of Ecology."

Very briefly let us review the situation in the light of present facts and hazard a few guesses about the future.

On the biological level, advancing science and technology have set going a revolutionary process that seems to be destined for the next century at least, perhaps for much longer, to exercise a decisive influence upon the destinies of all human societies and their individual members. In the course of the last fifty years extremely effective methods for lowering the prevailing rates of infant and adult mortality were developed by Western scientists. These methods were very simple and could be applied with the expenditure of very little money by very small numbers of not very highly trained technicians. For these reasons, and because everyone regards life as intrinsically good and death as intrinsically bad, they were in fact applied on a worldwide scale. The results were spectacular. In the past, high birthrates were balanced by high death rates. Thanks to science, death rates have been halved but, except in the most highly industrialized, contraceptive-using countries, birthrates remain as high as ever. An enormous and accelerating increase in human numbers has been the inevitable consequence.

At the beginning of the Christian era, so demographers assure us, our planet supported a human population of about two hundred and fifty millions. When the Pilgrim fathers stepped ashore, the figure had risen to about five hundred millions. We see, then, that in the relatively recent past it took sixteen hundred years for the human species to double its numbers. Today world population stands at three thousand millions. By the year 2000, unless something appallingly bad or miraculously good should happen in the interval, six thousand millions of us will be sitting down to breakfast every morning. In a word, twelve times as many people are destined to double their numbers in one-fortieth of the time.

This is not the whole story. In many areas of the world human numbers are increasing at a rate much higher than the average for the whole species. In India, for example, the rate of increase is now 2.3 per cent per annum. By 1990 its four hundred and fifty million inhabitants will have become nine hundred million inhabitants. A comparable rate of increase will raise the population of China to the billion mark by 1980. In Ceylon, in Egypt, in many of the countries of South and Central America, human numbers are increasing at an annual rate of three per cent. The result will be a doubling of their present populations in about twenty-three years.

On the social, political, and economic levels, what is likely to happen in an underdeveloped country whose people double themselves in a single generation, or even less? An underdeveloped society is a society without adequate capital resources (for capital is what is left over after primary needs have been satisfied, and in underdeveloped countries most people never satisfy their primary needs); a society without a sufficient force of trained teachers, administrators, and technicians; a society with few or no industries and few or no developed sources of industrial power; a society, finally, with enormous arrears to be made good in food production, education, road building, housing, and sanitation. A quarter of a century from now, when there will be twice as many of them as there are today, what is the likelihood that the members of such a society will be better fed, housed, clothed, and schooled than at present? And what are the chances in such a society for the maintenance, if they already exist, or the creation, if they do not exist, of democratic institutions?

Mr. Eugene Black, the former president of the World Bank, once expressed the opinion that it would be extremely difficult, perhaps even impossible, for an underdeveloped country with a very rapid rate of population increase to achieve full industrialization. All its resources, he pointed out, would be absorbed year by year in the task of supplying, or not quite supplying, the primary needs of its new members. Merely to stand still, to maintain its current subhumanly inadequate standard of living, will require hard work and the expenditure of all the nation's available capital. Available capital may be increased by loans and gifts from abroad, but in a world where the industrialized nations are involved in power politics and an increasingly expensive armament race there will never be enough foreign aid to make much difference. And even if the loans and gifts to underdeveloped countries were to be substantially increased, any resulting gains would be largely nullified by the uncontrolled population explosion.

The situation of these nations with such rapidly increasing populations reminds one of Lewis Carroll's parable in *Through the Looking Glass*, where Alice and the Red Queen start running at full speed and run for a long time until Alice is completely out of breath. When they stop, Alice is amazed to see that they are still at their starting point. In the looking glass world, if you wish to retain your present position, you must run as fast as you can. If you wish to get ahead, you must run at least twice as fast as you can.

If Mr. Black is correct (and there are plenty of economists and demographers who share his opinion), the outlook for most of the world's newly independent and economically non-viable nations is gloomy indeed. To those that have shall be given. Within the next ten or twenty years, if war can be avoided, poverty will almost have disappeared from the highly industrialized and contraceptive-using societies of the West. Meanwhile, in the underdeveloped and uncontrolled breeding societies of Asia, Africa, and Latin America the condition of the masses (twice as numerous, a generation from now, as they are today) will have become no better and may even be decidedly worse than it is at present. Such a decline is foreshadowed by current statistics of the Food and Agriculture Organization of the United Nations. In some underdeveloped regions of the world, we are told, people are somewhat less adequately fed, clothed, and housed than were their parents and grandparents thirty and forty years ago. And what of elementary education? UNESCO provided an answer. Since the end of World War II heroic efforts have been made to teach the whole world how to read. The population explosion has largely stultified these efforts. The absolute number of illiterates is greater now than at any time.

The contraceptive revolution which, thanks to advancing science and technology, has made it possible for the highly developed societies of the West to offset the consequences of death control by a planned control of births, has had as yet no effect upon the family life of people in underdeveloped countries. This is not surprising. Death control, as I have already remarked, is easy, cheap, and can be carried out by a small force of technicians. Birth control, on the other hand, is rather expensive, involves the whole adult population, and demands of those who practice it a good deal of forethought and directed will power. To persuade hundreds of millions of men and women to abandon their tradition-hallowed views of sexual morality, then to distribute and teach them to make use of contraceptive devices or fertility-controlling drugs—this is a huge and difficult task, so huge and so difficult that it seems very unlikely that it can be successfully carried out, within a sufficiently short space of time, in any of the countries where control of the birthrate is most urgently needed.

Extreme poverty, when combined with ignorance, breeds that lack of desire for better things which has been called "wantlessness"—the resigned acceptance of a subhuman lot. But extreme poverty, when it is combined with the knowledge that some societies are affluent, breeds envious desires and the expectation that these desires must of necessity, and very soon, be satisfied. By means of the mass media (those easily exportable products of advancing science and technology) some knowledge of what life is like in affluent societies has been widely disseminated throughout the world's underdeveloped regions. But, alas, the science and technology which have given the industrial West its cars, refrigerators, and contraceptives have given the people of Asia, Africa, and Latin America only movies and radio broadcasts, which they are too simpleminded to be able to criticize, together with a population explosion, which they are still too poor and too tradition-bound to be able to control through deliberate family planning.

In the context of a three, or even of a mere two per cent annual increase in numbers, high expectations are foredoomed to disappointment. From disappointment, through resentful frustration, to widespread social unrest, through chaos, to dictatorship, possibly of the Communist Party, more probably of generals and colonels. It would seem, then, that for two-thirds of the human race now suffering from the consequences of uncontrolled breeding in a context of industrial backwardness, poverty, and illiteracy, the prospects for democracy, during the next ten or twenty years, are very poor.

From underdeveloped societies and the probable political consequences of their explosive increase in numbers we now pass to the prospect for democracy in the fully industrialized, contraceptive-using societies of Europe and North America.

It used to be assumed that political freedom was a necessary precondition of scientific research. Ideological dogmatism and dictatorial institutions were supposed to be incompatible with the open-mindedness and the freedom of experimental action, in the absence of which discovery and invention are impossible. Recent history has proved these comforting assumptions to be completely unfounded. It was under Stalin that Russian scientists developed the A-bomb and, a few years later, the H-bomb. And it is under a more-than-Stalinist dictatorship that Chinese scientists are now in the process of performing the same feat.

Another disquieting lesson of recent history is that, in a developing society, science and technology can be used exclusively for the enhancement of military power, not at all for the benefit of the masses. Russia has demonstrated, and China is now doing its best to demonstrate, that poverty and primitive conditions of life for the overwhelming majority of the population are perfectly compatible with the wholesale production of the most advanced and sophisticated military hardware. Indeed, it is by deliberately imposing poverty on the masses that the rulers of developing industrial nations are able to create the capital necessary for building an armament industry and maintaining a well-equipped army with which to play their parts in the suicidal game of international power politics.

We see, then, that democratic institutions and libertarian traditions are not at all necessary to the progress of science and technology, and that such progress does not of itself make for human betterment at home and peace abroad. Only where democratic institutions already exist, only where the masses can vote their rulers out of office and so compel them to pay attention to the popular will, are science and technology used for the benefit of the majority as well as for increasing the power of the state. Most human beings prefer peace to war, and practically all of them would rather be alive than dead. But in every part of the world men and women have been brought up to regard nationalism as axiomatic and war between nations as something cosmically ordained by the Nature of Things. Prisoners of their culture, the masses, even when they are free to vote, are inhibited by the fundamental postulates of the frame of reference within which they do their thinking and their feeling from decreeing an end to the collective paranoia that governs international relations. As for the world's ruling minorities, by the very fact of their power they are chained even more closely to the current system of ideas and the prevailing political customs; for this reason they are even less capable than their subjects of expressing the simple human preference for life and peace.

Some day, let us hope, rulers and ruled will break out of the cultural prison in which they are now confined. Some day. And may that day come soon! For, thanks to our rapidly advancing science and technology, we have very little time at our disposal. The river of change flows ever faster, and somewhere downstream, perhaps only a few years ahead, we shall come to the rapids, shall hear, louder and ever louder, the roaring of a cataract.

Modern war is a product of advancing science and technology. Conversely, advancing science and technology are products of modern war. It was in order to wage war more effectively that first the United States, then Britain and the U.S.S.R., financed the crash programs that resulted so quickly in the harnessing of atomic forces. Again, it was primarily for military purposes that the techniques of automation, which are now in the process of revolutionizing industrial production and the whole system of administrative and bureaucratic control, were initially developed. "During World War II," writes Mr. John Diebold, "the theory and use of feedback was studied in great detail by a number of scientists both in this country and in Britain. The introduction of rapidly moving aircraft very quickly made traditional gun-laying techniques of anti-aircraft warfare obsolete. As a result, a large part of scientific manpower in this country was directed toward the development of self-regulating devices and systems to control our military equipment. It is out of this work that the technology of automation as we understand it today has developed."

The headlong rapidity with which scientific and technological changes, with all their disturbing consequences in the fields of politics and social relations, are taking place is due in large measure to the fact that, both in the U.S.A. and the U.S.S.R., research in pure and applied science is lavishly financed by military planners whose first concern is in the development of bigger and better weapons in the shortest possible time. In the frantic effort, on one side of the Iron Curtain, to keep up with the Joneses—on the other, to keep up with the Ivanovs—these military planners spend gigantic sums on research and development. The military revolution advances under forced draft, and as it goes forward it initiates an uninterrupted succession of industrial, social, and political revolutions. It is against this background of chronic upheaval that the members of a species, biologically and historically adapted to a slowly changing environment, must now live out their bewildered lives.

Old-fashioned war was incompatible, while it was being waged, with democracy. Nuclear war, if it is ever waged, will prove in all likelihood to be incompatible with civilization, perhaps with human survival. Meanwhile, what of all the preparations for nuclear war? If certain physicists and military planners had their way, democracy, where it exists, would be replaced by a system of regimentation centered upon the bomb shelter. The entire population would have to be systematically drilled in the ticklish operation of going underground at a moment's notice, systematically exercised in the art of living troglodytically under conditions resembling those in the hold of an eighteenth-century slave ship. The notion fills most of us with horror. But if we fail to break out of the ideological prison of our nationalistic and militaristic culture, we may find ourselves compelled by the military consequences of our science and technology to descend into the steel and concrete dungeons of total and totalitarian civil defense.

In the past, one of the most effective guarantees of liberty was governmental inefficiency. The spirit of tyranny was always willing, but its technical and organizational flesh was weak. Today the flesh is as strong as the spirit. Governmental organization is a fine art, based upon scientific principles and disposing of marvelously efficient equipment. Fifty years ago an armed revolution still had some chance of success. In the context of modern weaponry a popular uprising is foredoomed. Crowds armed with rifles and homemade grenades are no match for tanks. And it is not only to its armament that a modern government owes its overwhelming power. It also possesses the strength of superior knowledge derived from its communication systems, its stores of accumulated data, its batteries of computers, its network of inspection and administration.

Where democratic institutions exist and the masses can vote their rulers out of office, the enormous powers with which science, technology, and the arts of organization have endowed the ruling minority are used with discretion and a decent regard for civil and political liberty. Where the masses can exercise no control over their rulers, these powers are used without compunction to enforce ideological orthodoxy and to strengthen the dictatorial state. The nature of science and technology is such that it is peculiarly easy for a dictatorial government to use them for its own antidemocratic purposes. Well-financed, equipped, and organized, an astonishingly small number of scientists and technologists can achieve prodigious results. The crash program that produced the A-bomb and ushered in a new historical era was planned and directed by some four thousand theoreticians, experimenters, and engineers. To parody the words of Winston Churchill, never have so many been so completely at the mercy of so few.

Throughout the nineteenth century the state was relatively feeble, and its interest in, and influence upon, scientific research were negligible. In our day the state is everywhere exceedingly powerful and a lavish patron of basic and ad-hoc research. In western Europe and North America the relations between the state and its scientists on the one hand and individual citizens, professional organizations, and industrial, commercial, and educational institutions on the other are fairly satisfactory. Advancing science, the population explosion, the armament race, and the steady increase and centralization of political and economic power are still compatible, in countries that have a libertarian tradition, with democratic forms of government. To maintain this compatibility in a rapidly changing world, bearing less and less resemblance to the world in which these democratic institutions were developed—this, quite obviously, is going to be increasingly difficult.

A rapid and accelerating population increase that will nullify the best efforts of underdeveloped societies to better their lot and will keep two-thirds of the human race in a condition of misery in anarchy or of misery under dictatorship, and the intensive preparations for a new kind of war that, if it breaks out, may bring irretrievable ruin to the one-third of the human race now living prosperously in highly industrialized societies—these are the two main threats to democracy now confronting us. Can these threats be eliminated? Or, if not eliminated, at least reduced?

My own view is that only by shifting our collective attention from the merely political to the basic biological aspects of the human situation can we hope to mitigate and shorten the time of troubles into which, it would seem, we are now moving. We cannot do without politics; but we can no longer afford to indulge in bad, unrealistic politics. To work for the survival of the species as a whole and for the actualization in the greatest possible number of individual men and women of their potentialities for good will, intelligence, and creativity—this, in the world of today, is good, realistic politics. To cultivate the religion of idolatrous nationalism, to subordinate the interests of the species and its individual members to the interests of a single national state and its ruling minority—in the context of the population explosion, missiles, and atomic warheads, this is bad and thoroughly unrealistic politics. Unfortunately, it is to bad and unrealistic politics that our rulers are now committed.

Ecology is the science of the mutual relations of organisms with their environment and with one another. Only when we get it into our collective head that the basic problem confronting twentieth-century man is an ecological problem will our politics improve and become realistic. How does the human race propose to survive and, if possible, improve the lot and the intrinsic quality of its individual members? Do we propose to live on this planet in symbiotic harmony with our environment? Or, preferring to be wantonly stupid, shall we choose to live like murderous and suicidal parasites that kill their host and so destroy themselves?

Committing that sin of overweening bumptiousness which the Greeks called *hubris*, we behave as though we were not members of earth's ecological community, as though we were privileged and, in some sort, supernatural beings and could throw our weight around like gods. But in fact we are, among other things, animals—emergent parts of the natural order. If our politicians were realists, they would think rather less about missiles and the problem of landing astronauts on the moon, rather more about hunger and moral squalor and the problem of enabling three billion men, women, and children, who will soon be six billions, to lead a tolerably human existence without, in the process, ruining and befouling their planetary environment.

Animals have no souls; therefore, according to the most authoritative Christian theologians, they may be treated as though they were things. The truth, as we are now beginning to realize, is that even things ought not to be treated as *mere* things. They should be treated as though they were parts of a vast living organism. "Do as you would be done by." The Golden Rule applies to our dealings with nature no less than to our dealings with our fellow men. If we hope to be well treated by nature, we must stop talking about "mere things" and start treating our planet with intelligence and consideration.

Power politics in the context of nationalism raises problems that, except by war, are practically insoluble. The problems of ecology, on the other hand, admit of a rational solution and can be tackled without the arousal of those violent passions always associated with dogmatic ideology and nationalistic idolatry. There may be arguments about the best way of raising wheat in a cold climate or of reforesting a denuded mountain. But such arguments never lead to organized slaughter. Organized slaughter is the result of arguments about such questions as: Which is the best nation? The best religion? The best political theory? The best form of government? Why are other people so stupid and wicked? Why can't they see how good and intelligent *we* are? Why do they resist our beneficent efforts to bring them under our control and make them like ourselves?

To questions of this kind the final answer has always been war. "War," said Clausewitz, "is not merely a political act but also a political instrument, a continuation of political relationships, a carrying out of the same by other means." This was true enough in the eighteen-thirties, when Clausewitz published his famous treatise, and it continued to be true until 1945. Now, obviously, nuclear weapons, long-range rockets, nerve gases, bacterial aerosols, and the laser (that highly promising addition to the world's military arsenals) have given the lie to Clausewitz. All-out war with modern weapons is no longer a continuation of previous policy; it is a complete and irreversible break with previous policy.

Power politics, nationalism, and dogmatic ideology are luxuries that the human race can no longer afford. Nor, as a species, can we afford the luxury of ignoring man's ecological situation. By shifting our attention from the now completely irrelevant and anachronistic politics of nationalism and military power to the problems of the human species and the still inchoate politics of human ecology we shall be killing two birds with one stone—reducing the threat of sudden destruction by scientific war and at the same time reducing the threat of more gradual biological disaster.

The beginnings of ecological politics are to be found in the special services of the United Nations Organization. UNESCO, the Food and Agriculture Organization, the World Health Organization, the various technical-aid services—all these are, partially or completely, concerned with the ecological problems of the human species. In a world where political problems are thought of and worked upon within a frame of reference whose coordinates are nationalism and military power, these ecology-oriented organizations are regarded as peripheral. If the problems of humanity could be thought about and acted upon within a frame of reference that has survival for the species, the well-being of individuals, and the actualization of man's desirable potentialities as its coordinates, these peripheral organizations would become central. The subordinate politics of survival, happiness, and personal fulfillment would take the place now occupied by the politics of power, ideology, nationalistic idolatry, and unrelieved misery.

In the process of reaching this kind of politics we shall find, no doubt, that we have done something, in President Wilson's prematurely optimistic words, "to make the world safe for democracy."

REFERENCES AND NOTES

• "ECOLOGY—THE RELENTLESS SCIENCE"

[1]The process of interaction in the environment is traced and analyzed thoroughly in Part II of the monumental report of the International Symposium at Princeton entitled *Man's Role in Changing the Face of the Earth*, edited by William L. Thomas, Jr. (Chicago: University of Chicago Press, 1956). Among the major topics are such headings as the modifications of biotic communities, the alterations of climatic elements, and the ecology of wastes.

[2]David Cushman Coyle, *The Ordeal of the Presidency* (Washington: Public Affairs Press, 1960), p. 208.

[3]A descriptive discourse on ecological relationships is that of Philip L. Wagner, *The Human Use of the Earth* (Glencoe, Ill.: The Free Press, 1960), especially Chapter 3.

Check Jack McCormick, *The Living Forest* (New York: Harper & Bros., 1959), for an analysis of the deciduous forest community.

George K. Reid, *Ecology of Land Waters and Estuaries* (New York: Reinhold Publishing Corp., 1961), is a source that is rich.

A highly informative text appropriate for the serious lay reader is that of Herbert C. Hanson and Ethan D. Churchill, *The Plant Community* (New York: Reinhold Publishing Corp., 1961).

Of a different sort and very thought-provoking is Joseph Harold Rush's *The Dawn of Life* (New York: Doubleday & Co., 1958), in which he suggests that man, by recklessly meddling with natural selection and natural processes, may have effectively checkmated his own evolution and mental and physical progress.

[4]*Science*, Mar. 2, 1962, pp. 709ff.

[5]A scholarly account through the "corridors of time" of man's stewardship, or lack of it, is found on pages 115ff. of the report of the International Symposium, *Man's Role in Changing the Face of the Earth*, cited above.

[6]Tom Gill, "Forest Pest Control," *Control and Wildlife Relationships: A Symposium*, Publication 897, National Academy of Sciences—National Research Council, Washington, D.C., 1961, p. 11.

[7]*Science*, Mar. 2, 1962, p. 711.

[8]Robert C. Cowen, "Why the Spider Waits in Vain," *Christian Science Monitor*, Dec. 1, 1964.

[9]Marian Sorenson, "Hawks Help Man," *Christian Science Monitor*, May 2, 1964.

[10]See R. MacDonald, "Death Road for the Condor," *Sports Illustrated*, Apr., 1964, pp. 39ff. Among a dozen such pleas for this raptor in recent popular journals.

[11]Marshall T. Case, "Dead Trees—Vital Links in a Chain of Life," *Cornell Plantations*, XVIII, 2, Summer, 1962.

[12]Report of a study by Dr. William A. Niering, Dr. R. H. Whittaker, and Dr. C. H. Lowe headlined "Extinction Perils Giant U.S. Cactus," in the *Christian Science Monitor*, Jan. 18, 1965, written by Ruby Zagoren.

[13]"If We Slaughter a Bird . . . ," a review of *Our Animal Neighbors* by Alan Devoe with Mary Berry Devoe (New York: McGraw-Hill), in *National Humane Review*, June, 1963, p. 12.

[14]Quoted by Malcolm Cowley in "The Writer as Craftsman," *Saturday Review*, June 27, 1964, p. 18.

[15]*Sierra Club Bulletin*, Sept., 1965.

[16]"Is Civilization Progress?" *Izaak Walton Magazine*, Nov., 1964.

[17]Governor Rolvaag had the courage to veto 1965 bounty bills passed by the legislature. Minnesota has been practicing the wasteful system of paying bounties in the amount of $301,000 from fish and game funds each year. This money is now to be used by the counties for wildlife and habitat development—the positive rather than the negative approach in the "new conservation." Wisconsin's Conservation Department Director L. D. Voigt hails the state's recent elimination of the bounty system and declares that its yearly cost of $180,000 may now be put toward "constructive fish and game projects—the most significant step forward . . . since the creation of public hunting and fishing grounds."

[18]The Secretary of the Interior seeks an amendment to the Criminal Code to extend protective federal laws for the benefit of wildlife; among the benefactors would be the harassed Florida alligators, who "lose their hides" to poachers at the rate of $1 million a year. Women buying alligator bags and shoes unwittingly foster this racket.

[19]*Man and Nature in America* (New York: Columbia University Press, 1965) gives a valuable picture incorporating a larger landscape.

THE HUMAN ECOSYSTEM

[1]K. Boulding, *Human Values on the Spaceship Earth* (New York: National Council of Churches of Christ in the U.S.A., 1966), p. 14.

[2]V. G. Childe, *Man Makes Himself* (London: Watts & Co., 1941).

[3]C. P. Snow, *The Two Cultures and the Scientific Revolution* (New York: Cambridge University Press, 1959).

[4]C. Elton, *Animal Ecology* (New York: Macmillan, 1936).

[5]T. R. Malthus, *An Essay on the Principle of Population As It Affects the Future Improvement of Mankind*, 1798. (Facsimile reprint in 1926 for J. Johnson. London: Macmillan & Co.).

[6]P. S. Dawson, "Xenocide, Suicide and Cannibalism in Flour Beetles," *American Naturalist* 102(924): 97-105.

[7]L. B. Slobodkin, *Growth and Regulation of Animal Populations* (New York: Holt, Rinehart & Winston, 1962).

L. B. Slobodkin, "Aspects of the Future of Ecology," *BioScience* 18(1): 16-23.

L. B. Slobodkin, F. E. Smith, and N. G. Hairston, "Regulation in Terrestrial Ecosystems, and the Implied Balance of Nature," *American Naturalist* 101: 109-124.

[8]P. R. Ehrlich, "World Population: A Battle Lost?" *Stanford Today* ser. 1., no. 22, p. 4.

[9]Elton, *op. cit.*

[10]M. Bates, *The Prevalence of People* (New York: Scribner, 1955).

[11]K. Boulding, *Human Values on the Spaceship Earth* (New York: National Council of Churches of Christ in the U.S.A., 1966).

● POPULATION POLICY: WILL CURRENT PROGRAMS SUCCEED?

[1]*Studies in Family Planning, No. 16* (1967).

[2]*Ibid No. 9* (1966), p. 1.

[3]The statement is given in *Studies in Family Planning (1*, p. 1), and in *Population Bull.* **23**, 6 (1967).

[4]The statement is quoted in *Studies in Family Planning (1*, p. 2).

[5]*Hearings on S. 1676, U.S. Senate, Subcommittee on Foreign Aid Expenditures, 89th Congress, Second Session, April 7, 8, 11* (1966), pt. 4, p. 889.

[6]B. L. Raina, in *Family Planning and Population Programs*, B. Berelson, R. K. Anderson, O. Harkavy, G. Maier, W. P. Mauldin, S. G. Segal, Eds. (Univ. of Chicago Press, Chicago, 1966).

[7]D. Kirk, *Ann. Amer. Acad. Polit. Soc. Sci.* **369**, 53 (1967).

[8]As used by English-speaking demographers, the word *fertility* designates actual reproductive performance, not a theoretical capacity.

[9]K. Davis, *Rotarian* **94**, 10 (1959); *Health Educ. Monographs* **9**, 2 (1960); L. Day and A. Day, *Too Many Americans* (Houghton Mifflin, Boston, 1964); R. A. Piddington, *Limits of Mankind* (Wright, Bristol, England, 1956).

[10]*Official Gazette* (15 Apr. 1965); quoted in *Studies in Family Planning (1*, p. 7).

[11]J. W. Gardner, Secretary of Health, Education, and Welfare, "Memorandum to Heads of Operating Agencies" (Jan. 1966), reproduced in *Hearings on S. 1676 (5)*, p. 783.

[12]C. Tietze, *Demography* **1**, 119 (1964); *J. Chronic Diseases* **18**, 1161 (1964); M. Muramatsu, *Milbank Mem. Fund Quart.* **38**, 153 (1960); K. Davis, *Population Index* **29**, 345 (1963); R. Armijo and T. Monreal, *J. Sex Res.* **1964**, 143 (1964); Proceedings World Population Conference, Belgrade, 1965; Proceedings International Planned Parenthood Federation.

[13]*Studies in Family Planning, No. 4* (1964), p. 3.

[14]D. Bell (then administrator for Agency for International Development), in *Hearings on S. 1676 (5)*, p. 862.

[15] *Asian Population Conference* (United Nations, New York, 1964), p. 30.

[16] R. Armijo and T. Monreal, in *Components of Population Change in Latin America* (Milbank Fund, New York, 1965), p. 272; E. Rice-Wray, *Amer. J. Public Health* **54**, 313 (1964).

[17] E. Rice-Wray, in "Intra-Uterine Contraceptive Devices," *Excerpta Med. Intern. Congr. Ser. No. 54* (1962), p. 135.

[18] J. Blake, in *Public Health and Population Change*, M. C. Sheps and J. C. Ridley, Eds. (Univ. of Pittsburgh Press, Pittsburgh, 1965), p. 41.

[19] J. Blake and K. Davis, *Amer. Behavioral Scientist*, **5**, 24 (1963).

[20] See "Panel discussion on comparative acceptability of different methods of contraception," in *Research in Family Planning*, C. V. Kiser, Ed. (Princeton Univ. Press, Princeton, 1962), pp. 373-86.

[21] "From the point of view of the woman concerned, the whole problem of continuing motivation disappears, . ." [D. Kirk, in *Population Dynamics*, M. Muramatsu and P. A. Harper, Eds. (Johns Hopkins Press, Baltimore, 1965)].

[22] "For influencing family size norms, certainly the examples and statements of public figures are of great significance . . . also . . . use of mass-communication methods which help to legitimize the small-family style, to provoke conversation, and to establish a vocabulary for discussion of family planning." [M. W. Freymann, in *Population Dynamics*, M. Muramatsu and P. A. Harper, Eds. (Johns Hopkins Press, Baltimore, 1965)].

[23] O. A. Collver, *Birth Rates in Latin America* (International Population and Urban Research, Berkeley, Calif., 1965), pp. 27-28; the ten countries were Colombia, Costa Rica, El Salvador, Ecuador, Guatemala, Honduras, Mexico, Panamà, Peru, and Venezuela.

[24] J. R. Rele, *Fertility Analysis through Extension of Stable Population Concepts*. (International Population and Urban Research, Berkeley, Calif., 1967).

[25] J. C. Ridley, M. C. Sheps, J. W. Lingner, J. A. Menken, *Milbank Mem. Fund Quart.* **45**, 77 (1967); E. Arriaga, unpublished paper.

[26] "South Korea and Taiwan appear successfully to have checked population growth by the use of intrauterine contraceptive devices" [U. Borell, *Hearings on S. 1676 (5)*, p. 556].

[27] K. Davis, *Population Index* **29**, 345 (1963).

[28] R. Freedman, *ibid.* **31**, 421 (1965).

[29] Before 1964 the Family Planning Association had given advice to fewer than 60,000 wives in 10 years and a Pre-Pregnancy Health Program had reached some 10,000 and, in the current campaign, 3650 IUD's were inserted in 1965, in a total population of 2 1/2 million women of reproductive age. See *Studies in Family Planning, No. 19* (1967), p. 4, and R. Freedman *et al., Population Studies* **16**, 231 (1963).

[30] R. W. Gillespie, *Family Planning on Taiwan* (Population Council, Taichung, 1965), p. 45.

[31]During the period 1950-60 the ratio of growth of the city to growth of the noncity population was 5:3; during the period 1960-64 the ratio was 5:2; these ratios are based on data of Shaohsing Chen, *J. Sociol. Taiwan* **1**, 74 (1963) and data in the United Nations *Demographic Yearbooks*.

[32]Gillespie, *op. cit.*, p. 69.

[33]R. Freedman, *Population Index* **31**, 434 (1965). Taichung's rate of decline in 1963-64 was roughly double the average in four other cities, whereas just prior to the campaign its rate of decline had been much less than theirs.

[34]Gillespie, *op. cit.*, pp. 18, 31.

[35]*Ibid*, p. 8.

[36]S. H. Chen, *J. Soc. Sci. Taipei* **13**, 72 (1963).

[37]R. Freedman *et al., Population Studies* **16**, 227 (1963); *ibid.*, p. 232.

[38]In 1964 the life expectancy at birth was already 66 years in Taiwan, as compared to 70 for the United States.

[39]J. Blake, *Eugenics Quart.* **14**, 68 (1967).

[40]Women accepting IUD's in the family-planning program are typically 30 to 34 years old and have already had four children. [*Studies in Family Planning No. 19* (1967), p. 5].

[41]Y. K. Cha, in *Family Planning and Population Programs*, B. Berelson *et al.*, Eds. (Univ. of Chicago Press, Chicago, 1966), p. 27.

[42]*Ibid*, p. 25.

[43]H. S. Ayalvi and S. S. Johl, *J. Family Welfare* **12**, 60 (1965).

[44]Sixty percent of the women had borne their first child before age 19. Early marriage is strongly supported by public opinion. Of couples polled in the Punjab, 48 percent said that girls *should* marry before age 16, and 94 percent said they should marry before age 20 (H. S. Ayalvi and S. S. Johl, *ibid.*, p. 57). A study of 2380 couples in 60 villages of Uttar Pradesh found that the women had consummated their marriage at an average age of 14.6 years [J. R. Rele, *Population Studies* **15**, 268 (1962)].

[45]J. Morsa, in *Family Planning and Population Programs*, B. Berelson *et al.*, Eds. (Univ. of Chicago Press, Chicago, 1966).

[46]H. Gille and R. J. Pardoko, *ibid.*, p. 515; S. N. Agarwala, *Med. Dig. Bombay* **4**, 653 (1961).

[47]*Mysore Population Study* (United Nations, New York, 1961), p. 140.

[48]A. Daly, in *Family Planning and Population Programs*, B. Berelson *et al.*, Eds. (Univ. of Chicago Press, Chicago, 1966).

[49]*Mysore Population Study* (United Nations, New York, 1961), p. 140.

[50]C. J. Goméz, paper presented at the World Population Conference, Belgrade, 1965.

[51]C. Miro, in *Family Planning and Population Programs*, B. Berelson *et al.*, Eds. (Univ. of Chicago Press, Chicago, 1966).

[52]*Demographic Training and Research Centre (India) Newsletter* **20**, 4 (Aug. 1966).

[53]K. Davis, *Population Index* **29**, 345 (1963). For economic and sociological theory of motivation for having children, see J. Blake [Univ. of California (Berkeley)], in preparation.

[54]K. Davis, *Amer. Economic Rev.* **46**, 305 (1956); *Sci. Amer.* **209**, 68 (1963).

[55]J. Blake, *World Population Conference [Belgrade, 1965]* (United Nations, New York, 1967), vol. 2, pp. 132-36.

[56]S. Enke, *Economics Statistics* **42**, 175 (1960); ——, *Econ. Develop. Cult. Change* **8**, 339 (1960); ——, *ibid.* **10**, 427 (1962); A. O. Krueger and L. A. Sjaastad, *ibid.*, p. 423.

[57]T. J. Samuel, *J. Family Welfare India* **13**, 12 (1966).

[58]Sixty-two countries, including 27 in Europe, give cash payments to people for having children [U.S. Social Security Administration, *Social Security Programs Throughout the World, 1967* (Government Printing Office, Washington, D.C., 1967), pp. xxvii-xxviii].

[59]Average gross reproduction rates in the early 1960's were as follows: Hungary, 0.91; Bulgaria, 1.09; Romania, 1.15; Yugoslavia, 1.32.

[60]Blake, *op. cit.*, p. 1195.

[61]O. A. Collver and E. Langlois, *Econ. Develop. Cult. Change* **10**, 367 (1962); J. Weeks, [Univ. of California (Berkeley)], unpublished paper.

[62]Roman Catholic textbooks condemn the "small" family (one with fewer than four children) as being abnormal [J. Blake, *Population Studies* **20**, 27 (1966)].

[63]Judith Blake's critical readings and discussions have greatly helped in the preparation of this article.

THE TRAGEDY OF THE COMMONS

[1]J. B. Wiesner and H. F. York, *Sci. Amer.* **211** (No. 4), 27 (1964).

[2]G. Hardin, *J. Hered.* **50**, 68 (1959); S. von Hoernor, *Science* **137**, 18 (1962).

[3]J. von Neumann and O. Morgenstern, *Theory of Games and Economic Behavior* (Princeton Univ. Press, Princeton, N.J., 1947), p. 11.

[4]J. H. Fremlin, *New Sci.*, No. 415 (1964), p. 285.

[5]A. Smith, *The Wealth of Nations* (Modern Library, New York, 1937), p. 423.

[6]W. F. Lloyd, *Two Lectures on the Checks to Population* (Oxford Univ. Press, Oxford, England, 1833), reprinted (in part) in *Population, Evolution, and Birth Control*, G. Hardin, Ed. (Freeman, San Francisco, 1964), p. 37.

[7]A. N. Whitehead, *Science and the Modern World* (Mentor, New York, 1948), p. 17.

[8]G. Hardin, Ed. *Population, Evolution, and Birth Control* (Freeman, San Francisco, 1964), p. 56.

[9]S. McVay, *Sci. Amer.* **216** (No. 8), 13 (1966).

[10]J. Fletcher, *Situation Ethics* (Westminster, Philadelphia, 1966).

[11]D. Lack, *The Natural Regulation of Animal Numbers* (Clarendon Press, Oxford, 1954).

[12]H. Girvetz, *From Wealth to Welfare* (Stanford Univ. Press, Stanford, Calif., 1950).

[13]G. Hardin, *Perspec. Biol. Med.* **6**, 366 (1963).

[14]U. Thant, *Int. Planned Parenthood News*, No. 168 (February 1968), p. 3.

[15]K. Davis, *Science* **158**, 730 (1967).

[16]S. Tax, Ed., *Evolution after Darwin* (Univ. of Chicago Press, Chicago, 1960), vol. 2, p. 469.

[17]G. Bateson, D. D. Jackson, J. Haley, J. Weakland, *Behav. Sci.* **1**, 251 (1956).

[18]P. Goodman, *New York Rev. Books* **10**(8), 22 (23 May 1968).

[19]A. Comfort, *The Anxiety Makers* (Nelson, London, 1967).

[20]C. Frankel, *The Case for Modern Man* (Harper, New York, 1955), p. 203.

[21]J. D. Roslansky, *Genetics and the Future of Man* (Appleton-Century-Crofts, New York, 1966), p. 177.

● THE OUTLYING ROCKS

[1]Epigraph. Quoted in Joel A. Allen.

[2]Aldo Leonard, *A Sand County Almanac*.

[3]"Saga of Eric the Red," in *America: A Library of Original Sources* Vol. 1 (Chicago: Veterans of Foreign Wars, 1925).

[4]In *America: A Library of Original Sources* Vol. 1 (Chicago: Veterans of Foreign Wars, 1925).

[5]*Labrador Journal*, quoted in Fisher and Lockley.

[6]*Manual of the Ornithology of the United States and Canada*.

[7]Journals of Columbus, quoted in Welker.

[8]The accounts of this shipwreck are said to have inspired Shakespeare's *The Tempest*.

[9]Quoted in Bent, *Life Histories of . . . Petrels. . . .*

[10]J. B. Labat, *Nouveau Voyage aux Isles de l'Amerique*, 1722 (*ibid.*).

(For full information about incomplete references see the bibliography in *Wildlife in America*.).

TO SURVIVE ON THE EARTH

[1]See "The Supersonic Transport," by Kurt H. Hohenemser, in *Scientist and Citizen*, April 1966.

[2]For an illuminating discussion of the significance of the technological failure of the auto industry, see the article "The Murderous Motor Car" by Lewis Mumford in the *New York Review of Books* for April 28, 1966.

[3]For a discussion of the urgency of transforming urban waste-removal systems, see *Waste Management and Control*, a report to the Federal Council for Science and Technology by the Committee on Pollution (Washington, D. C.: publication 1400 of the National Research Council, National Academy of Sciences, 1966). Among other things the report points out that at the present rate of accumulation of pollutants, essentially all of the surface waters of the United States will become so contaminated as to lose their biological capability for purification within the next twenty years.

[4]Although it is apparent that runoff of fertilizers from farm land makes a growing contribution to the pollution of surface water by nitrate and phosphate, especially the former, detailed information about this problem is surprisingly scarce. As in the case of all pollution problems, little can be done about it until we really know the sources of specific pollutants. There is an urgent need for detailed study of agricultural fertilizer practices and their impact on the pollution of surface water.

THE ECONOMICS OF THE COMING SPACESHIP EARTH

[1]Ludwig von Bertalanffy, *Problems of Life* (New York: John Wiley and Sons, 1952).

[2]Richard L. Meier, *Science and Economic Development* (New York: John Wiley and Sons, 1956).

[3]K. E. Boulding, "The Consumption Concept in Economic Theory," *American Economic Review*, 35:2 (May 1945), pp. 1-14; and "Income or Welfare?", *Review of Economic Studies*, 17 (1949-50) pp. 77-86.

[4]Fred L. Polak, *The Image of the Future*, Vols. I and II, translated by Elise Boulding (New York: Sythoff, Leyden and Oceana, 1961).

BIBLIOGRAPHY

Here is a list of books of interest to the general reader on the subjects presented in this book. These sources will, in turn, suggest additional reading for those interested in going further into the field. The following periodicals also frequently carry articles on environmental issues: *Audubon, Environment, Natural History, Scientific American*.

Annual Reports. Resources for the Future, Inc.: 1755 Massachusetts Avenue, N.W., Washington, D.C. 20036

Blake, Peter. *God's Own Junkyard*. Holt, Rinehart & Winston: New York, 1964.

Brown, Harrison S. *The Challenge of Man's Future*. Viking Press: New York, 1954.

Calder, Ritchie. *Living with the Atom*. University of Chicago Press: Chicago, Illinois, 1962.

Carson, Rachel. *Silent Spring*. Houghton Mifflin: Boston, Massachusetts, 1962.

Commoner, Barry. *Science and Survival*. Viking Press: New York, 1967.

Crowe, Philip Kingsland. *The Empty Ark*. Charles Scribner's Sons: New York, 1967.

Curtis, Richard, and Elizabeth Hogan. *Perils of the Peaceful Atom: The Myth of Safe Nuclear Power Plants*. Ballantine: New York, 1969.

Darling, F. Fraser, and John P. Milton, eds. *Future Environments of North America*. Natural History Press: Garden City, New York, 1966.

Dasmann, Raymond F. *A Different Kind of Country*. Macmillan: New York, 1968.

De Bell, Garrett, ed. *The Environmental Handbook*. Ballantine: New York, 1970.

Douglas, William O. *A Wilderness Bill of Rights*. Little, Brown and Company: Boston, 1965.

Dubos, René. *So Human An Animal*. Charles Scribner's Sons: New York, 1968.

Eckardt, Wolf von. *A Place to Live: The Crisis of the Cities.* Delacorte Press: New York, 1967.

Ehrlich, Paul R. *The Population Bomb.* Ballantine: New York, 1968.

Ehrlich, Paul R., and Anne H. Ehrlich. *Population, Resources, Environment.* W. H. Freeman: San Francisco, 1970.

Goldman, Marshall I., ed. *Controlling Pollution: The Economics of a Cleaner America.* Prentice-Hall: Englewood Cliffs, New Jersey, 1967.

Graham, Frank, Jr. *Disaster by Default: Politics and Water Pollution.* M. Evans: New York, 1966.

Graham, Frank, Jr. *Since Silent Spring.* Houghton Mifflin Company: Boston, Massachusetts, 1970.

Hardin, Garrett, ed. *Population, Evolution and Birth Control.* W. H. Freeman: San Francisco, 1969.

Herber, Lewis. *Our Synthetic Environment.* Alfred A. Knopf: New York, 1962.

Herfindahl, Orris C., and Allen V. Kneese. *Quality of the Environment: An Economic Approach to Some Problems in Using Land, Water, and Air.* Resources for the Future: Washington, D.C., 1965.

Huxley, Sir Julian. *The Human Crisis.* University of Washington Press: Seattle, Washington, 1963.

Jarrett, Henry, ed. *Environmental Quality in a Growing Economy.* Johns Hopkins Press: Baltimore, Maryland, 1966.

Leopold, Aldo. *A Sand County Almanac.* Oxford University Press: New York, 1949.

Marine, Gene. *America the Raped.* Simon & Schuster: New York, 1969.

Matthiessen, Peter. *Wildlife in America.* Viking Press: New York, 1964.

Mumford, Lewis. *The City in History: Its Origins, Its Transformations, Its Prospects.* Harcourt, Brace & World: New York, 1961.

Nash, Roderick, ed. *The American Environment: Readings in the History of Conservation.* Addison-Wesley: Reading, Massachusetts, 1968.

Novick, Sheldon. *The Careless Atom.* Houghton Mifflin Company: Boston, 1969.

Osborn, Fairfield. *Our Plundered Planet.* Little, Brown and Company: Boston, 1948.

Paddock, William, and Paul Paddock. *Famine—1975.* Little, Brown and Company: Boston, 1967.

Perry, John. *Our Polluted World: Can Man Survive?* Franklin Watts: New York, 1967.

Restoring the Quality of Our Environment. U.S. President's Advisory Committee, Environmental Pollution Panel: The White House, November, 1965.

Rienow, Robert, and Leona Train Rienow. *Moment in the Sun.* Dial Press: New York, 1967.

Rudd, Robert L. *Pesticides and the Living Landscape.* University of Wisconsin Press: Madison, Wisconsin, 1964.

Thoreau, Henry David. *Walden.* Boston, 1854.

Tupper, Margo. *No Place to Play.* Chilton: Philadelphia, 1966.

Udall, Stewart. *1976: Agenda for Tomorrow.* Harcourt, Brace & World: New York, 1968.

Udall, Stewart. *The Quiet Crisis.* Holt, Rinehart & Winston: New York, 1963.

A 0
B 1
C 2
D 3
E 4
F 5
G 6
H 7
I 8
J 9